普通高等学校“十三五”省级规划教材

普通高等学校城市轨道交通专业规划教材

信息技术专业英语

Professional English for Information Technology

主　编　魏化永　陆晓君

副主编　周　岩　王绍陇

编写人员（以姓氏笔画为序）

王绍陇　王莹莹　刘　燕

吴天琪　吴红云　杨玉菡

陆晓君　周　岩　甄　扬

魏化永

主　审　李　锐

中国科学技术大学出版社

内 容 简 介

本书是编者多年从事高校信息技术专业英语课程教学和研究的经验总结，结合计算机类专业和英语教学的特点，旨在培养高校计算机类专业学生基于工作岗位的英语应用能力，努力提升学生可持续发展的职业核心竞争力。本书选材广泛，内容包括计算机及计算机系统、软件和应用、数据和数据库、通信与网络、大数据与云计算、人工智能和机器学习、计算机程序设计、计算机安全和道德、软件调试与测试等计算机基础知识。每章设置了阅读、写作、职业岗位、字词短语和练习五个部分，由浅入深，从基本理论到新技术、新概念逐步展开。

本书可作为高校计算机专业相关课程的教材，也可供从业人员自学参考。

图书在版编目(CIP)数据

信息技术专业英语/魏化永，陆晓君主编. —合肥：中国科学技术大学出版社，2022.7
(普通高等学校城市轨道交通专业规划教材)
ISBN 978-7-312-05455-6

Ⅰ. 信… Ⅱ. ①魏… ②陆… Ⅲ. 电子计算机—英语 Ⅳ. TP3

中国版本图书馆 CIP 数据核字(2022)第 101780 号

信息技术专业英语
XINGXI JISHU ZHUANYE YINGYU

出版 中国科学技术大学出版社
安徽省合肥市金寨路 96 号，230026
http://press.ustc.edu.cn
https://zgkxjsdxcbs.tmall.com
印刷 安徽省瑞隆印务有限公司
发行 中国科学技术大学出版社
开本 787 mm×1092 mm 1/16
印张 12
字数 378 千
版次 2022 年 7 月第 1 版
印次 2022 年 7 月第 1 次印刷
定价 40.00 元

Foreword 序

本套教材根据城市轨道交通运营管理、城市轨道交通通信信号技术、城市轨道交通车辆技术、城市轨道交通机电技术、城市轨道交通供配电技术专业的人才培养需要，结合对职业岗位能力的要求，由安徽交通职业技术学院、南京铁道职业技术学院、郑州铁路职业技术学院、上海工程技术大学、沈阳交通高等专科学校、新疆交通职业技术学院、合肥职业技术学院、合肥铁路工程学校、合肥市轨道交通集团有限公司、深圳城市轨道交通运营公司、杭州城市轨道交通运营公司、宁波城市轨道交通运营公司、郑州铁路局等单位共同编写。

本套教材整合了国内主要城市轨道交通运营企业现场作业的内容，以实际工作项目为主线，以项目中的具体工作任务作为知识学习要点，并针对各项任务设计模拟实训与思考练习，实现了通过课堂环境模拟现场岗位作业情景促进学生自我学习、自我训练的目标，体现了“岗位导向、学练一体”的教学理念。

本套教材涵盖城市轨道交通运营管理、城市轨道交通通信信号技术、城市轨道交通车辆技术、城市轨道交通机电技术、城市轨道交通供配电技术专业，可作为以上各相关专业课程的教材，并可供相关城市轨道交通运营企业相关人员参考。

普通高等学校城市轨道交通专业规划教材
编写委员会

Preface
前言

在全球经济一体化迅速发展的今天，英语作为国际通用语言，使用愈发广泛。随着数字经济时代的到来，英语已经全面深入计算机领域的教学和应用之中。计算机的很多基础操作使用的都是英文指令，对信息技术从业人员的英语水平提出了较高要求。

信息技术专业英语是计算机及相关信息技术专业的一门重要基础课程，其主要任务是培养学生实际运用英语的能力。通过本课程的学习，使学生掌握必备的信息技术专业词汇、术语、基本概念和相关应用技能，可以借助工具阅读信息技术专业书籍并能进行信息技术专业英语资料的翻译，为更好地使用计算机打下坚实的基础。

本书充分考虑了高校学生学习和就业需要，突出以下特色：

1. 系统实用的选材内容

切合学生实际，内容注重时代性、系统性与实用性相结合，既能提高学生的语言能力，又有利于培养学生的职业素养与技能。取材于真实的工作场景，以此来提高学生的学习兴趣。

2. 模块设计的教学内容

教学内容采用模块化设计，突出“工学结合、能力为本”的理念，每个单元浓缩为一个典型工作环节，学习任务与工作目标协同，实现“教、学、做”一体化。

3. 任务驱动的编写模式

以任务为基本单元，由浅入深、循序渐进，知识和技能螺旋式地融入于任务中，选材难度适当，每章配有关键词及相应习题，嵌入岗位能力描述，以培养学生的职业能力。

本书由安徽交通职业技术学院魏化永(第1章、第2章)和陆晓君(第4章、附录Ⅱ)担任主编，由安徽交通职业技术学院周岩(第5章)和王绍陇(第6章)担任副主编，安徽交通职业技术学院吴红云(第9章)、王莹莹(第3章)、杨玉菡(附录Ⅰ)、吴天琪(第7章)和安徽财贸职业学院刘燕(第8章)、安徽工业经济职业技术学院甄扬(第10章)参与编写，阿德莱德大学研究生吴悦然参与书稿整理工作，全书由安徽交通职业技术学院李锐主审。

在本书的编写过程中，编者参考了国内外出版社出版的一些教材和专著，借鉴了相关网站的相应内容等，在此对相关文献作者一并致以诚挚的谢意！

由于编者水平有限，缺点和错误在所难免，恳请读者批评指正！

编　者

2022年2月

Contents
目录

Chapter 1 Computers and computer systems

Learning objectives

After you have read this chapter, you should be able to:

☆ Explain the computer and its main components.
☆ Compare different computers, including PC, laptop, mainframe and server.
☆ Describe the differences between input and output devices.
☆ Identify the input devices and output devices.
☆ Explain how to get an Email and write letter.
☆ Describe Microprocessor, including control unit and the arithmetic-logic unit.
☆ Explain the hardware system of computer, including CPU, memory, I/O and bus lines.

Computers have become a vital part of everyday life. It is almost inconceivable that you could spend a day without at least one event being influenced by a computer. Perhaps the word "computer" automatically conjures up the image of a personal computer sitting on a desk, but in fact it is the computers you cannot see that influence your life the most. Typical examples of common products that may use these "invisible" computers are: cars, washing machines, bar-code reading systems, central-heating controllers, microwave ovens, etc.

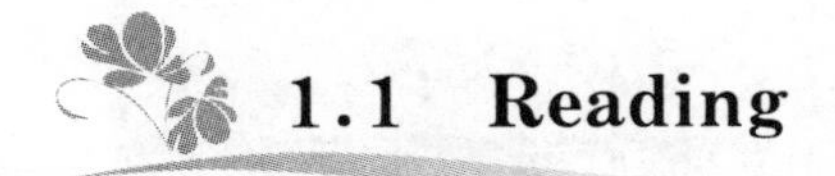

1.1 Reading

Passage 1: What is a computer

A computer is a machine that manipulates data following a list of instructions that

have been programmed into it. A programmable machine that performs high-speed processing of numbers, as well as of text, graphics, symbols, and sound. All computers contain a Central Processing Unit (CPU) that interprets and executes instructions; input devices, such as a keyboard and a mouse, through which data and commands into the computer; memory that enables the computer to store programs and data; and output devices, such as printers and display screens, that show the results after the computer has processed data.

The term "PC" is originally used to describe an IBM-compatible Personal Computer (PC), which has an Intel CPU as its microprocessor. Basically, it would run a typical Windows Operating System (OS). Other types of machines that are often not considered as PCs are Macs and servers running UNIX or Linux. A great number of companies produce PCs, and the PC is the most popular type of computer on the market. It comes in the form of a desktop computer (See Figure 1.1), which include separate components such as a case, a monitor, a keyboard, a mouse, and speakers; or in the form of a laptop computer (See Figure 1.2), which has all those features in one unit.

Figure 1.1　Desktop computer

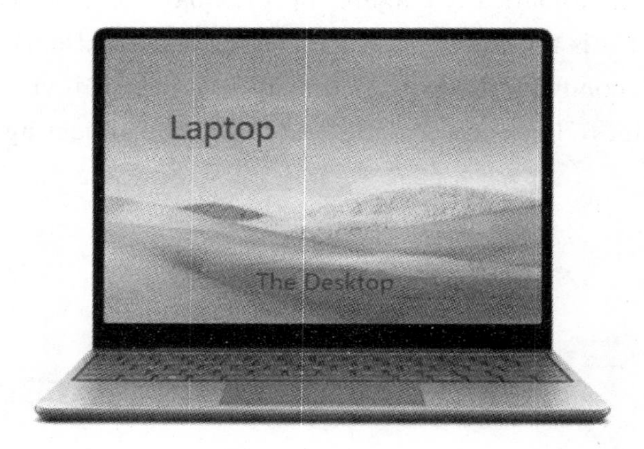

Figure 1.2　Laptop

The PC makes using the computer easier, especially with the Windows environment. This easy-to-use interface(so called GUI: Graphic User Interface) has made it possible for novice computer users to figure out how to use the computer.

Microprocessor: In most personal computer systems, the CPU or processor is contained on a single chip called the microprocessor (See Figure 1.3). The microprocessor is the "brains" of the computer system. It has two basic components: the control unit and the arithmetic-logic unit.

Figure 1.3 Intel microprocessor

Control unit: The control unit tells the rest of the computer system how to carry out a program's instructions. It directs the movement of electronic signals between memory, which temporarily holds data, instructions, processed information, and the arithmetic-logic unit. It also directs these control signals between the CPU and input and output devices.

Arithmetic-logic unit: The arithmetic-logic unit, usually called the ALU, performs two types of operations: arithmetic and logical. Arithmetic operations are the fundamental math operations: addition, subtraction, multiplication, and division. Logical operations consist of comparisons such as whether one item is equal to (=), less than (<), or greater than (>) the other.

Passage 2: Computer system

Most computer systems, from the embedded controllers found in automobiles and consumer appliances to personal computers and mainframes (See Figure 1.4), have the same basic organization. It includes hardware system and software system.

The hardware system has three main components: the CPU, the memory subsystem, and the Input/Output (I/O) subsystem. The generic organization of these components is shown in Figure 1.5. Physically, a bus is a set of wires. The components of the computer are connected to the buses. To send information from one component to another, the source component outputs data onto the bus. The destination component then inputs this data from the bus. The system shown in Figure 1.5 has three buses. The uppermost bus in the figure is the address bus. When the CPU reads data or instructions from or writes data to memory, it must specify the address of the memory location it wishes to access. When accessing an I/O device, the CPU places the address of the device on the address bus. Each device can read the address of the bus and determine whether it is the device being accessed by the CPU. Unlike the other buses, the address bus always receives data from the CPU; the CPU never reads the address bus.

(a) Huawei ACU2　　(b) Zhaoyang S5100

(c) China Shenwei Taihu Lake Light

Figure 1.4

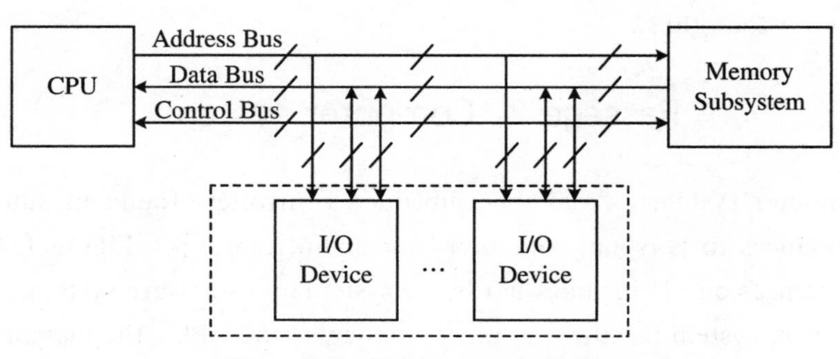

Figure 1.5　Computer organization

1.2 Writing: Email basics

Email basics

Do you ever feel like the only person who doesn't use Electronic mail (Email for short, see Figure 1.6)? You don't have to feel left out. If you're just getting started, you'll see that with a little bit of practice, email is easy to understand and use.

Figure 1.6 Email

In this section, you will learn what email is, how it compares to traditional mail, and how email addresses are written. We'll also discuss various types of email providers and the features and tools they include with an email account.

1. Getting to know email

Email is a way to send and receive messages across the Internet. It's similar to traditional mail (See Figure 1.7), but it also has some key differences. To get a better idea of what email is all about, take a look at the bellowing table (See Table 1.1) and consider how you might benefit from its use.

Figure 1.7 Traditional mail

Table 1.1 Traditional mail VS Email

Compare items	Traditional mail	Email
Address	Traditional mail is addressed with the recipient's name, street address, city, state or province, and zip code. For example, it will usually look something like this: Taihu Road 22, Hefei City, Anhui Province of China, 230051.	Email address are always written in a standard format, but they look quite different from traditional mail. An email address includes: a username, the @(at) symbol, and the email provider's domain. Username often include numbers and shortened versions of a name to create unique email address, and will usually look something like this: admin8@126.com.
Delivery	Traditional mail in a sealed envelope or package is delivered to a home or post office box by a mail carrier.	Email is delivered electronically across the internet. It is received by the **inbox** of an email service provider like QQ, Yahoo or 163.
Time	Traditional mail delivery could take anywhere between a couple of days, to a couple of weeks, depending on where it's being sent.	Email is delivered instantly, or usually within a few minutes.

2. Email advantages

Productivity tools: Email is usually packaged with a calendar, address book, instant messaging, and more for convenience and productivity.

- **Access to web services**: If you want to sign up for an account like Oracle or order products from services like Amazon, you will need an email address so you can be safely identified and contacted.
- **Easy mail management**: Email service providers have tools that allow you to file, label, prioritize, find, group, and filter your emails for easy management. You can even easily control spam or junk email.
- **Privacy**: Your email is delivered to your own personal and private account with a password required to access and view emails.
- **Communication with multiple people**: You can send an email to multiple people at once, giving you the option to include as few as or as many people as you want in a conversation.
- **Accessible anywhere at any time**: You don't have to be at home to get your mail. You can access it from any computer or mobile device that has an Internet connection.

3. Understanding email addresses

To receive emails, you will need an email account and an email address. Also, if you want to send emails to other people, you will need to obtain their email addresses. It's important to learn how to write email addresses correctly because if you do not enter them exactly right, your emails will not be delivered or might be delivered to the wrong person.

Email addresses are always written in a standard format that includes a username (See Figure 1.8), the @(at) symbol, and the email provider's domain.

The username is the name you choose to identify yourself.

Designing an Effective SQL Data Lakehouse ☆
From: KDnuggets <editor1@kdnuggets.com>
(Send by behalf of bounce-mc.us12_50554633.323097-0fc8927298@mail117.atl111.rsgsv.net)
Date: Monday, Jan 31, 2022 5:19 PM
To: hywei <hyweisky@foxmail.com>

Figure 1.8 Email address — username

The email provider (See Figure 1.9) is the website that hosts your email account.

Designing an Effective SQL Data Lakehouse ☆
From: KDnuggets <editor1@kdnuggets.com>
(Send by behalf of bounce-mc.us12_50554633.323097-0fc8927298@mail117.atl111.rsgsv.net)
Date: Monday, Jan 31, 2022 5:19 PM
To: hywei <hyweisky@foxmail.com>

Figure 1.9 Email address — provider

Some businesses and organizations use email addresses with their own website domain (See Figure 1.10).

Figure 1.10 Customized email address

4. Web-mail providers

In the past, people usually received an email account from the same companies that provided their Internet access. For example, if QQ provided your Internet connection, you'd have an QQ email address. While this is still true for some people, today it's increasingly common to use a free web-based email service, also known as web-mail. Anyone can use these services, no matter who provides their Internet access.

Today, the top two web-mail providers (See Figure 1.11) are QQ and Net Ease (mail.163). These providers are popular because they allow you to access your email account from anywhere with an Internet connection. You can also access web-mail on your mobile device.

Figure 1.11 The top two web-mail providers of China

5. Other email providers

Many people also have an email address hosted by their company, school, or organization. These email addresses are usually for professional purposes. For example, the people who work for this website have email addresses that end with @163.com. If you are part of an organization that hosts your email, they'll show you how to access it.

Many hosted web domains end with a suffix other than ".com." Depending on the organization, your provider's domain might end with a suffix like ".gov" for government websites, ".edu" for schools, ".mil" for military branches, or ".org" for nonprofit organizations.

6. Email applications

Many companies and organizations use an email application, like Microsoft Outlook, for communicating and managing their email. This software can be used with any email provider but is most commonly used by organizations that host their own email.

7. Email productivity features

In addition to email access, web-mail providers offer various tools and features. These features are part of a productivity suite — a set of applications that help you work, communicate, and stay organized. The tools offered will vary by providers, but all major web-mail services offer the following features:

- Instant messaging, or chat, which lets you have text-based conversations with other users.
- An online address book (See Figure 1.12), where you can store contact information for the people you contact frequently.
- An online calendar (See Figure 1.13) to help organize your schedule and share it with others.

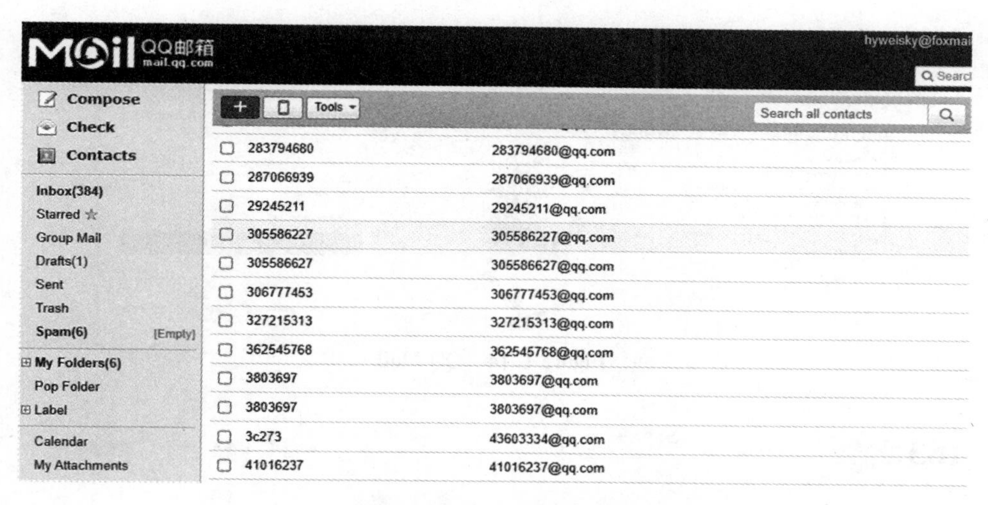

Figure 1.12 Address book

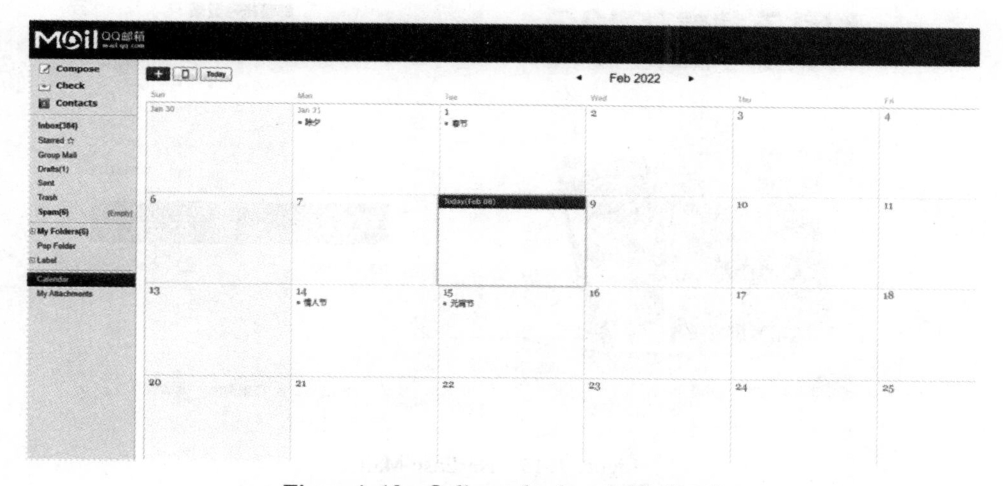

Figure 1.13 Online calendar of QQ Email

8. Getting started with email

You should now have a good understanding of what email is all about. Over the next few lessons, we will continue to cover essential email basics, etiquette, and safety tips.

9. Setting up your own email account

If you want to sign up for your own email account, we suggest choosing from one of the two major web-mail providers. But the simplest way is that as long as you have a QQ account, you have a QQ Email.

- QQ Mail: http://mail.qq.com (See Figure 1.14).
- NetEase Mail: https://mail.163.com (See Figure 1.15).

Figure 1.14　QQ Mail

Figure 1.15　NetEase Mail

10. Practice using an email program

Keep in mind that this tutorial will not show you how to use a specific email account. For that, you will need to visit QQ Email tutorial. It's a useful course for learning the basics, even if you ultimately end up choosing an email provider other than Gmail, such as Yahoo! or Outlook.com. There, you will learn how to:

- Sign up for an email account.
- Navigate and get to know the email interface.
- Compose, manage, and respond to email.
- Set up email on a mobile device.

Sample: Mail for help

Dear Professor Smith,

Could you please do me a favor at 10:00 a.m. on Monday, August 29? I am on a business trip, and it will be delayed until next month, so I really need you to run the monthly seminar in our department for me.

I know it might take some preparation time, but you are the only one I can count on, and I know that you are more than qualified to run this event. Please let me know if you can help.

I look forward to hearing from you and I hope I can return the favor sometime.

Sincerely yours,

David

1.3 Careers in IT

Computer support specialists

Computer support specialists provide technical support to customers and other users. They also may be called technical support specialists or help-desk technicians. Computer support specialists manage the every-day technical problems faced by computer users. They resolve common networking problems and may use troubleshooting programs to diagnose problems. Most computer support specialists are hired to work within a company and provide technical support for other employees and divisions. However, it is increasingly common for companies to provide technical support as an outsourced service.

Employers generally look for individuals with either an advanced associate's degree or a bachelor's degree to fill computer support specialist positions. Degrees in computer science or information systems may be preferred. However, because demand for qualified applicants is so high, those with practical experience and certification from a training program increasingly fill these positions. Employers seek individuals with customer service experience who demonstrate good analytical, communication, and people skills.

Computer technicians

Computer technicians repair and install computer components and systems. They may work on everything from personal computers and mainframe servers to printers. Some computer technicians are responsible for setting up and maintaining computer networks. Experienced computer technicians may work with computer engineers to diagnose problems and run routine maintenance on complex systems. Job growth is expected in this field as computer equipment becomes more complicated and technology expands.

Employers look for those with certification in computer repair or associate's degrees from professional schools. Computer technicians also can expect to continue their education to keep up with technological changes. Good communication skills are important in this field.

Opportunities for advancement typically come in the form of work on more advanced computer systems. Some computer technicians move into customer service positions or go into sales.

1.4 Words and phrases

access
address bus
arithmetic-logic unit
brain
Central Processing Unit (CPU)
component
computer
computer support specialist
computer technician
control bus
control unit
data bus
desktop
destination
diagnose
display screen

domain
email account
embedded controller
feature
figure out
Graphic User Interface (GUI)
hardware
I/O device
input devices
instant message
instruction
interface
keyboard
laptop
mainframes
manipulate
memory
microprocessor
monitor
mouse
novice
Operating System(OS)
opportunity
output devices
printer
processor
receive email
server
single chip
source
spam email
store
subsystem
suffix
unique mail address
Windows environment

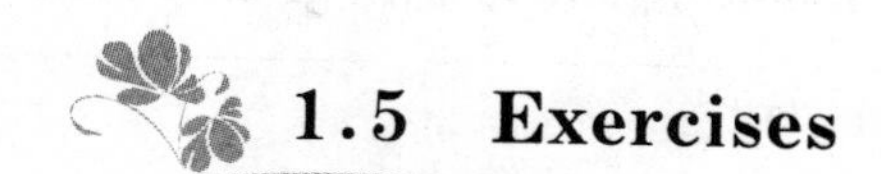

1.5 Exercises

Ⅰ. Matching

Match each numbered item with the most closely related lettered item. Write your answers in the spaces provided.

a. CPU
b. laptop
c. OS
d. interface
e. memory

________(1) the way a computer program presents information to a user or receives information from a user, in particular the layout of the screen and the menus; or an electrical circuit, connection or program that joins one device or system to another

________(2) any device internal to the computer, such as a primary hard disk drive or motherboard

________(3) an electronic, digital device that stores and processes information

________(4) central processing unit

________(5) integrated circuit semiconductor chip that performs the bulk of the processing and controls the parts of a system

f. component ________(6) a small computer that can work with a battery and be easily carried

g. computer ________(7) the device inner computer which store the data

h. microprocessor ________(8) operating system

i. novice ________(9) physical things that make up a computer, such as a component or a peripheral

j. hardware ________(10) a person who is new and has little experience in a skill, job or situation

Ⅱ. Written practice

1. What would you write at the end of a letter?

A. With love B. With best wishes C. Yours faithfully
D. Love to all E. Yours sincerely

(1) A letter for your boyfriend/girlfriend →( )
(2) A letter for your friends →( )
(3) A letter for your parents →( )
(4) A letter for just an acquaintance →( )
(5) If you don't know the name of the addressee →( )

2. Writing letters: Complete the following letters with given phrases.

A. Further to your request
B. I would be very grateful if you would
C. Following our telephone conversation
D. I just wanted to say thanks
E. Just a quick note

(6) ( ) for information regarding my house ...
(7) ( ) of this afternoon I'm writing to send you my Resume ...
(8) ( ) forward details of your interesting mail.
(9) ( ) to say Hello to you!
(10) ( ) for your marvelous gift.

Ⅲ. Open-ended questions

On a separate sheet of paper, respond to each question or statement.

(1) Explain the difference between PCs and specialized servers.
(2) Describe what I/O devices you know about computers.
(3) Explain the difference between RAM and ROM.

Chapter 2 Software and applications

Learning objectives

After you have read this chapter, you should be able to:

☆ Explain software, program and application.
☆ Identify general-purpose applications and specialized applications.
☆ Describe operating system, including definition, key functions of its.
☆ Compare different software, including system software, application software.
☆ Describe the boot processing of Linux.
☆ Describe how to customize your desktop background using image.

Computer software system (See Figure 2. 1) mainly includes system software and application software. Any program designed to run on a computer is called software. Application is a software program which allows a user to perform specific tasks such as

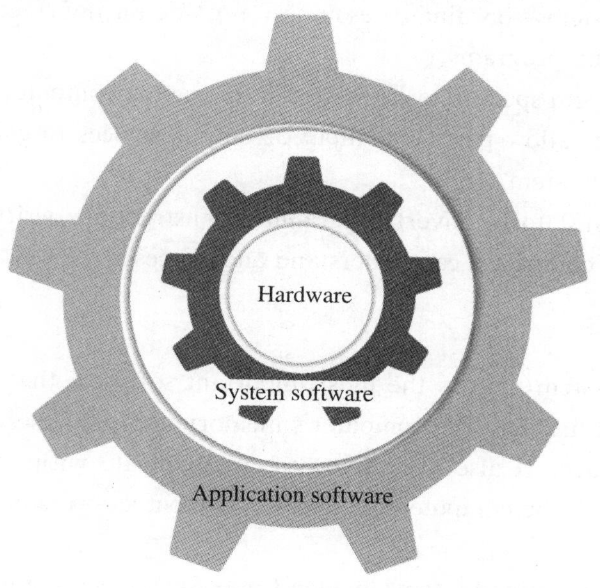

Figure 2. 1 Computer software system

word processing, email, accounting, database management.

"Examples of popular applications include Microsoft Words, QQ, WeChat, Tmall, Adobe Photoshop, Eclipse, Jupyter notebook, and Mozilla Firefox, etc."

2.1 Reading

There are two kinds of software. System software works with end users, application software, and computer hardware to handle the majority of technical details. Application software can be described as end user software and is used to accomplish a variety of tasks.

Passage 1: System software and operating system

Ⅰ. System software

System software works with end users, application programs, and computer hardware to handle many details relating to computer operations. It is not a single program but a collection or system of programs, these programs handle hundreds of technical details with little or no user intervention. Four kinds of systems programs are operating systems, utilities, device drivers, and language translators.

- Operating systems coordinate resources, provide an interface between users and the computer, and run programs.
- Utilities perform specific tasks related to managing computer resources.
- Device drivers allow particular input or output devices to communicate with the rest of the computer system.
- Language translators convert programming instructions written by programmers into a language that computers can understand and process.

Ⅱ. What is OS

An operating system(OS) is the most important software that runs on a computer (See Figure 2.2). It manages the computer's memory and processes, as well as all of its software and hardware. It also allows you to communicate with the computer without knowing how to speak the computer's language. Without an operating system, a computer is useless.

Some popular Operating Systems include Linux Operating System, Windows Operating System, VMS, Macs, Android, etc.

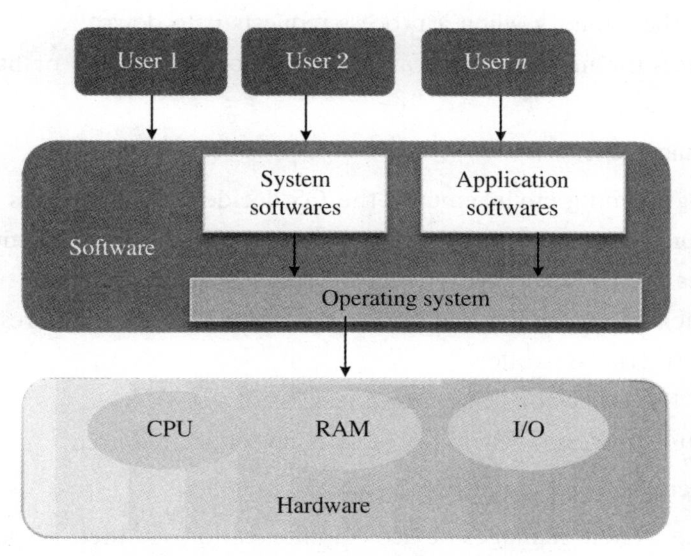

Figure 2. 2 Operating system architecture

OS is a program that acts as an interface and controls the execution of all kinds of programs.

Following are some of important functions of an operating system (See Table 2. 1).

Table 2. 1 Functions of OS

No.	Functions	No.	Functions
1	Memory management	6	Control over system performance
2	Processor management	7	Job accounting
3	Device management	8	Error detecting aids
4	File management	9	Coordination between other software and users
5	Security		

1. Memory management

Memory management refers to management of Primary Memory or Main Memory. Main memory is a large array of words or bytes where each word or byte has its own address.

Main memory provides a fast storage that can be accessed directly by the CPU. For a program to be executed, it must in the main memory. An Operating System does the following activities for memory management:

• Keeps tracks of primary memory, i. e., what part of it are in use by whom, what part are not in use.

• In multi-programming, the OS decides which process will get memory, when and how much.

• Allocates the memory when a process requests it to do so.

• De-allocates the memory when a process no longer needs it or has been terminated.

2. Processor management

In multi-programming environment, the OS decides which process gets the processor, when and for how much time. This function is called process scheduling. An Operating System does the following activities for processor management:

• Keeps tracks of processor and status of process. The program responsible for this task is known as traffic controller.

• Allocates the processor (CPU) to a process.

• De-allocates processor when a process is no longer required.

3. Device management

An Operating System manages device communication via their respective drivers. It does the following activities for device management:

• Keeps tracks of all devices. Program responsible for this task is known as the I/O controller.

• Decides which process gets the device, when and for how much time.

• Allocates the device in the efficient way.

• De-allocates devices.

4. File management

A file system is normally organized into directories for easy navigation and usage. These directories may contain files and other directories. An Operating System does the following activities for file management:

• Keeps track of information, location, uses, status etc. The collective facilities are often known as file system.

• Decides who gets the resources.

• Allocates the resources.

• De-allocates the resources.

5. Other important activities

Following are some of the important activities that an Operating System performs:

• Security — By means of password and similar other techniques, it prevents unauthorized access to programs and data.

• Control over system performance — Recording delays between request for a service and response from the system.

• Job accounting — Keeping track of time and resources used by various jobs and users.

• Error detecting aids — Production of dumps, traces, error messages, and other debugging and error detecting aids.

• Coordination between other software and users — Coordination and assignment of compilers, interpreters, assemblers and other software to the various users of the computer systems.

Program VS Software

Program: a set of instructions in code that control the operations or functions of a computer in order to perform a particular task.

Software: written programs or procedures or rules and associated documentations and the relevant data, pertaining to the operation of a computer system and that are stored in read/write memory.

Ⅲ. How to customize your desktop background?

Want your computer to feel a bit more like, well, your computer? You might want to consider changing your wallpaper.

The wallpaper is the image that appears behind the icons on your computer's desktop — that's why it's usually called a desktop background. On most computers, you can change your background by right-clicking the desktop and selecting personalize (See Figure 2. 3).

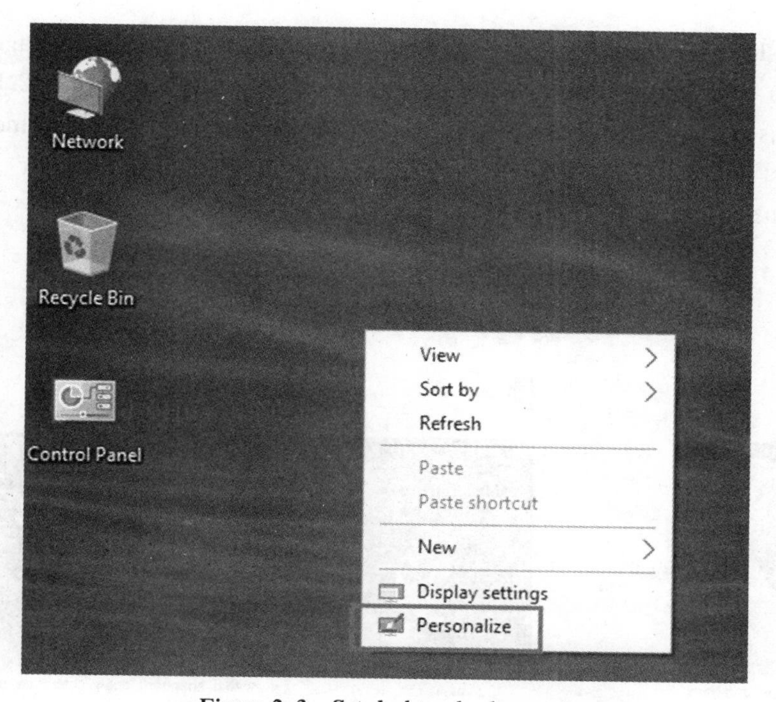

Figure 2. 3 Set desktop background

Then select desktop background. By default, you'll see the images that were included with your computer (See Figure 2.4).

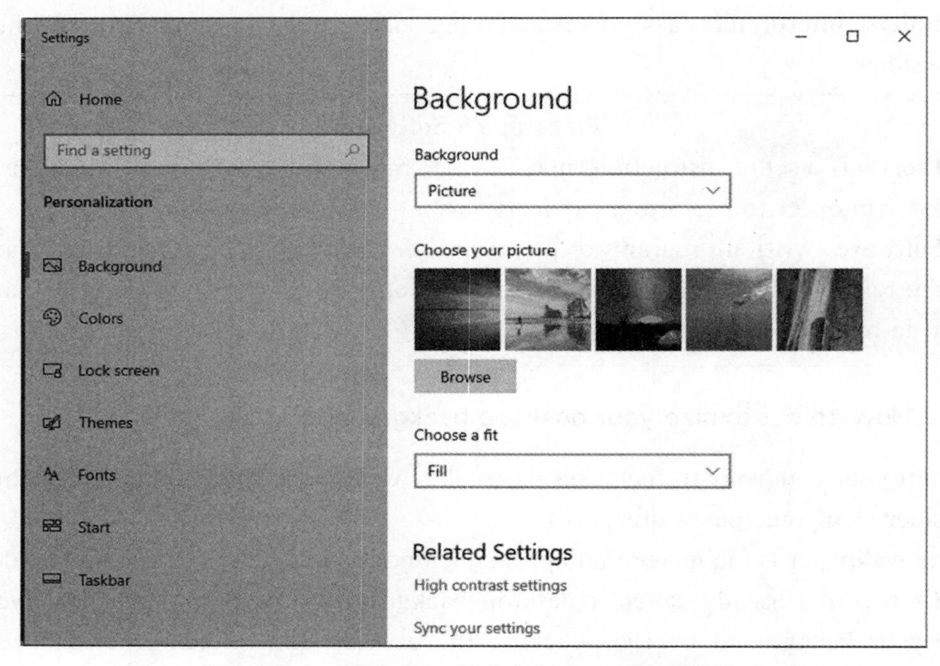

Figure 2.4 Set desktop background — select images

If you're looking for something specific, might we suggest Baidu Image Search? Its custom search tools allow you to search for images by size (See Figure 2.5 and Figure 2.6), which is perfect when you're looking for a large picture (we recommend an image that's at least 1920 × 1080). Anything smaller will look pixelated or grainy when

Figure 2.5 Searching the online images (1)

stretched across your desktop.

Figure 2.6 Searching the online images (2)

When you find an image you like, just click it, then you can open the links site to view it.

Then click and drag the picture to your desktop or click the download button to download it to your local disk (See Figure 2.7).

Figure 2.7 Download the online images to local

You can now set the image as your desktop background. Right-click the image and select Set as desktop background (See Figure 2.8 and Figure 2.9).

Figure 2. 8　Set the downloaded images as background

Figure 2. 9　The result of example

You can use this same technique to turn any of your personal photos into a desktop background.

Ⅳ. The Linux System boot processing

Boot: To start up a computer (See Figure 2.10).

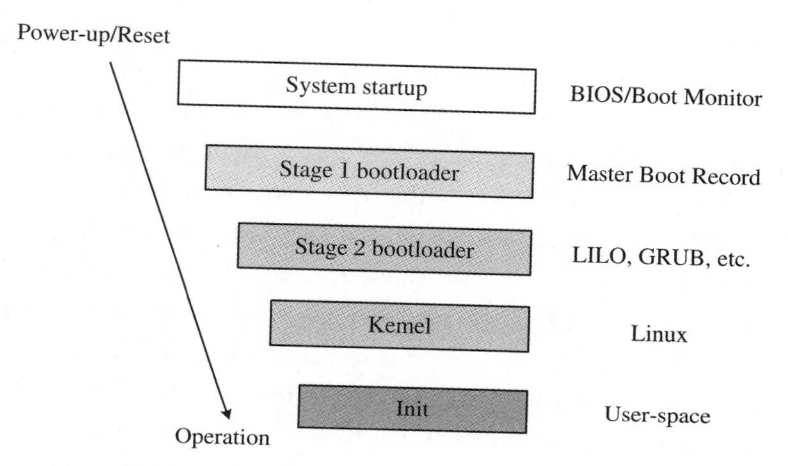

Figure 2.10 Steps of booting the Linux system

Step 1: POST

Starting an x86-based Linux system involves a number of steps. When the computer is powered on, the Basic Input/Output System (BIOS) initializes the hardware, including the screen and keyboard, and tests the main memory. This process is also called Power on Self-Test (POST).

The BIOS software is stored on a Read Only Memory (ROM) chip on the motherboard. After this, the remainder of the boot process is completely controlled by the operating system.

Step 2: Master Boot Records (MBR) and Boot Loader

Once the POST is completed, the system control passes from the BIOS to the boot loader. The boot loader is usually stored on one of the hard disks in the system, either in the boot sector (for traditional BIOS/MBR systems) or the EFI partition (for more recent Unified Extensible Firmware Interface or EFI/UEFI systems). Up to this stage, the machine does not access any mass storage media. Thereafter, information on the date, time, and the most important peripherals are loaded from the CMOS values (after a technology used for the battery-powered memory store — which allows the system to keep track of the date and time even when it is powered off).

A number of boot loaders exist for Linux; the most common ones are GRUB (for Grand Unified Boot loader) and ISO Linux (for booting from removable media).

Most Linux boot loaders can present a user interface for choosing alternative options for booting Linux, and even other operating systems that might be installed. When booting Linux, the boot loader is responsible for loading the kernel image and the initial RAM disk (which contains some critical files and device drivers needed to start the system) into memory.

Step 3: Boot Loader in action

The boot loader has two distinct stages:

First Stage: For systems using the BIOS/MBR method, the boot loader resides at the first sector of the hard disk also known as the Master Boot Record (MBR). The size of the MBR is just 512 bytes. In this stage, the boot loader examines the partition table and finds a bootable partition. Once it finds one, it then searches for the second stage boot loader, e.g., GRUB, and loads it into Random Access Memory (RAM).

For systems using the EFI/UEFI method, UEFI firmware reads its Boot Manager data to determine which UEFI application is to be launched and from where (i.e., from which disk and partition the EFI partition can be found). The firmware then launches the UEFI application, for example, GRUB, as defined in the boot entry in the firmware's boot manager. This procedure is more complicated but more versatile than the older MBR methods.

Second Stage: The second stage boot loader resides under/boot. A splash screen (See Figure 2.11) is displayed which allows us to choose which OS to boot. After choosing the OS, the boot loader loads the kernel of the selected operating system into RAM and passes control to it.

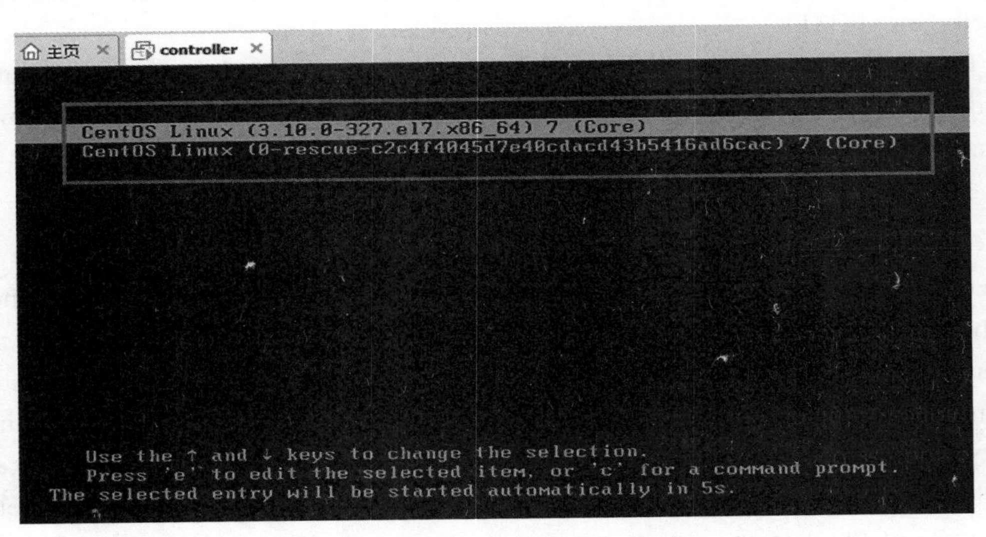

Figure 2.11 Splash screen of selection

The boot loader loads the selected kernel image (in the case of Linux) and passes control to it. Kernels are almost always compressed, so its first job is uncompressing itself. After this, it will check and analyze the system hardware and initialize any hardware device drivers built into the kernel.

Step 4: The Linux Kernel

The boot loader loads both the kernel and an initial RAM-based filesystem (initramfs) into memory so it can be used directly by the kernel.

When the kernel is loaded in RAM, it immediately initializes and configures the computer's memory and also configures all the hardware attached to the system. This includes all processors, I/O subsystems, storage devices, etc. The kernel also loads some necessary user space applications.

Step 5: The Initial RAM Disk

The initramfs filesystem image contains programs and binary files that perform all actions needed to mount the proper root filesystem, like providing kernel functionality for the needed filesystem and device drivers for mass storage controllers with a facility called udev (for User Device) which is responsible for figuring out which devices are present, locating the drivers they need to operate properly, and loading them. After the root filesystem has been found, it is checked for errors and mounted.

The mount program instructs the operating system that a filesystem is ready for use, and associates it with a particular point in the overall hierarchy of the filesystem (the mount point). If this is successful, the initramfs is cleared from RAM and the init program on the root filesystem (/sbin/init) is executed.

Init handles the mounting and pivoting over to the final real root filesystem. If special hardware drivers are needed before the mass storage can be accessed, they must be in the initramfs image.

Step 6: /sbin/init and Services

Once the kernel has set up all its hardware and mounted the root filesystem, the kernel runs the /sbin/init program. This then becomes the initial process, which then starts other processes to get the system running. Most other processes on the system trace their origin ultimately to init; the exceptions are kernel processes, started by the kernel directly for managing internal operating system details.

Traditionally, this process startup was done using conventions that date back to System V UNIX, with the system passing through a sequence of run-levels containing collections of scripts that start and stop services. Each run-level supports a different mode of running the system. Within each run-level, individual services can be set to run, or to be shut down if running. Newer distributions are moving away from the System V standard, but usually support the System V conventions for compatibility purposes.

Besides starting the system, init is responsible for keeping the system running and for shutting it down cleanly. It acts as the "manager of last resort" for all non-kernel processes, cleaning up after them when necessary, and restarts user login services as needed when users log in and out.

Step 7: Text-Mode Login

Near the end of the boot process, init starts a number of text-mode login prompts (done by a program called Getty, See Figure 2.12). These enable you to type your username, followed by your password, and to eventually get a command shell.

Usually, the default command shell is bash (the GNU Bourne Again Shell), but there are a number of other advanced command shells available. The shell prints a text prompt, indicating it is ready to accept commands; after the user types the command and presses Enter, the command is executed, and another prompt is displayed after the command is done.

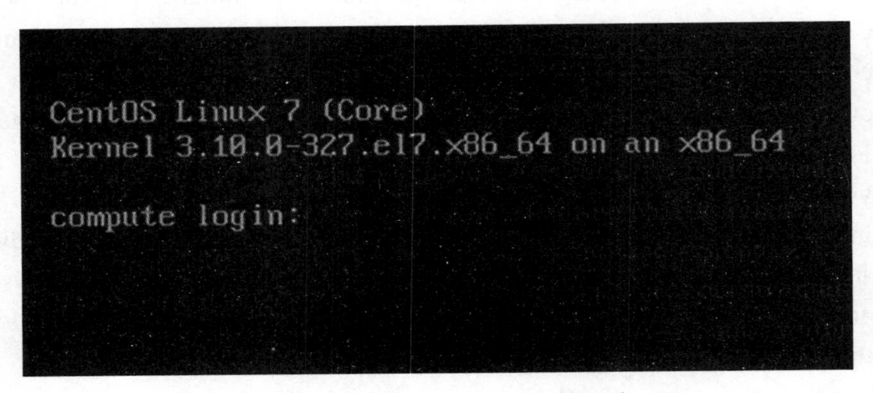

Figure 2.12 Text-mode — login interface

> **Tips: Cold boot VS Warm boot**
>
> **Cold boot** means restarting computer after the power is turned off.
>
> **Warm boot** means restarting computer without turning off the power.

Passage 2: Application software

Application software is one of the most commonly used software except the operating system when people use computers.

According to the host environment of the software, it can be divided into the following three categories. One category, desktop application software, include offices suites, database management systems, programming tool software, etc. Another category, web-based application software, which is mainly used by users to use software via browsers. For example, common portal websites, e-commerce websites, and some online tools. The third category, mobile application software (also known as Apps), is add-on features or programs typically designed for smartphones and tablets.

Of course, according to the use purpose of the software, it can be divided into the following three categories: general purpose Apps, specialized Apps (including thousands of other programs that tend to be more narrowly focused and used in specific disciplines and occupations) and mobile Apps.

Ⅰ. User interface: desktop Apps

User interface is the portion of the application that allows you to control and to interact with the program. Depending on the application, you can use a mouse, a pointer, a keyboard, and/or your voice to communicate with the application. Most general-purpose applications use a mouse and a graphical user interface (GUI) that displays graphical elements called icons to represent familiar objects. The mouse controls a pointer on the screen that is used to select items such as icons. Another feature is the use of windows to display information. A window is simply a rectangular area that can contain a document, program, or message. (Do not confuse the term window with the various versions of Microsoft's Windows operating systems, which are programs.) More than one window can be opened and displayed on the computer screen at one time.

Traditionally, most software programs use a system of menus, toolbars, and dialog boxes (See Figure 2.13).

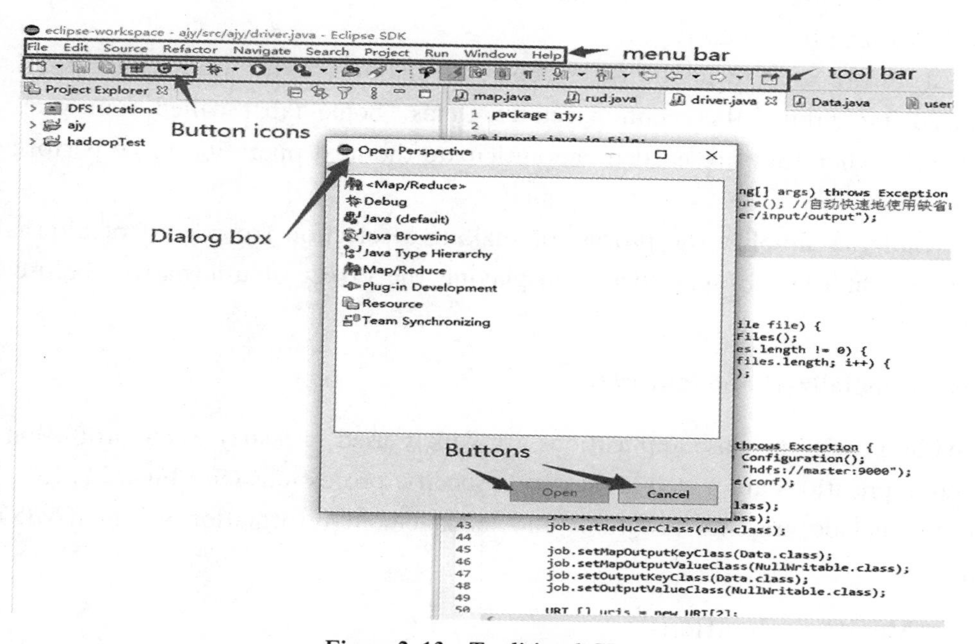

Figure 2.13 Traditional GUI

• Menus present commands that are typically displayed in a menu bar at the top of the screen.

• Toolbars typically appear below the menu bar and include small graphic elements called buttons that provide shortcuts for quick access to commonly used commands.

• Dialog boxes provide additional information and request user input. Many applications, and Microsoft applications in particular, use an interface known as the Ribbon GUI to make it easier to find and use all the features of an application; this GUI uses a system of ribbons, tabs, and galleries (See Figure 2. 14).

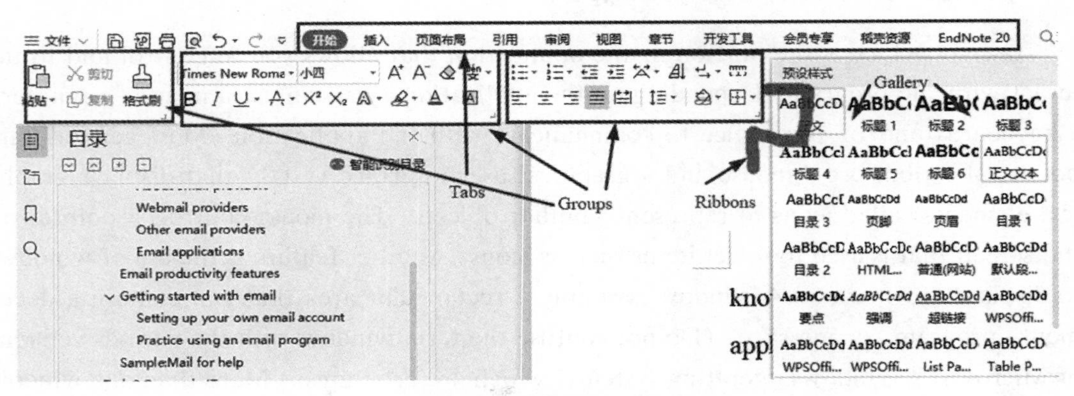

Figure 2. 14 Ribbon GUI

• Ribbons replace menus and toolbars by organizing commonly used commands into a set of tabs. These tabs display command buttons that are the most relevant to the tasks being performed by the user.

• Tabs are used to divide the ribbon into major activity areas. Each tab is then organized into groups that contain related items. Some tabs, called contextual tabs, appear only when they are needed and anticipate the next operation to be performed by the user.

• Galleries simplify the process of making a selection from a list of alternatives. This is accomplished by graphically displaying the effect of alternatives before being selected.

Ⅱ. Specialized applications

While general purpose applications are widely used in nearly every profession, specialized applications are widely used within specific professions (See Figure 2. 15). These programs include graphics programs and management information system (MIS) programs, etc.

Ⅲ. Mobile applications

Mobile Apps or mobile applications are add-on programs for a variety of mobile devices including smartphones and tablets (See Figure 2. 16). Sometimes referred to simply as Apps, mobile Apps have been widely used for years. The traditional applications include address books, to-do lists, alarms, and message lists. With the introduction of smartphones, tablets, and wireless connections to the Internet, mobile capabilities have exploded. Now, any number of applications are available. Mobile applications involve all areas of people's life, such as music, video, communication, shopping, games and so on.

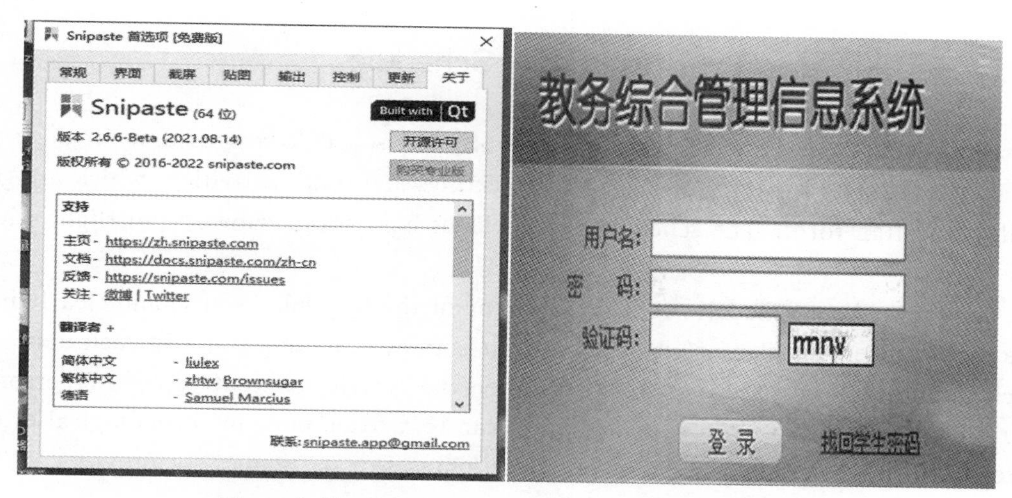

Figure 2. 15 Snipaste for capture and MIS for teaching

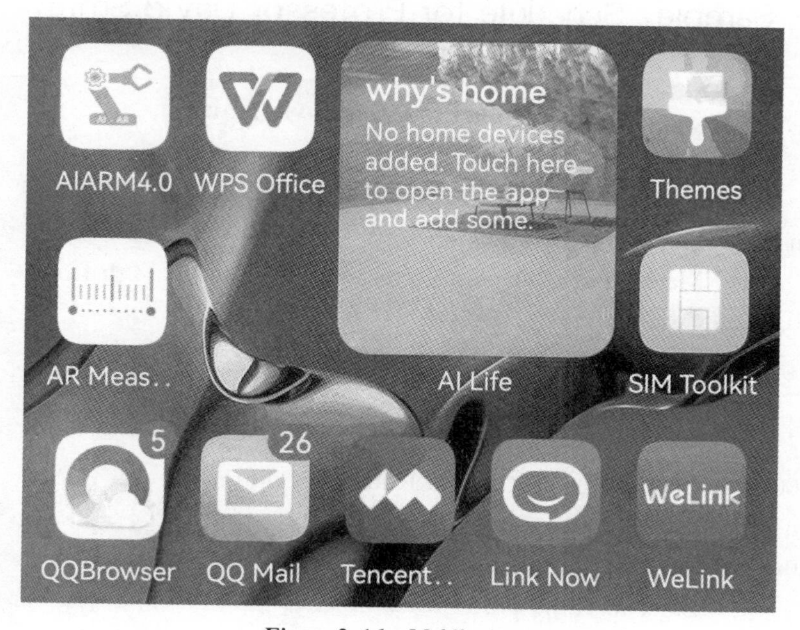

Figure 2. 16 Mobile Apps

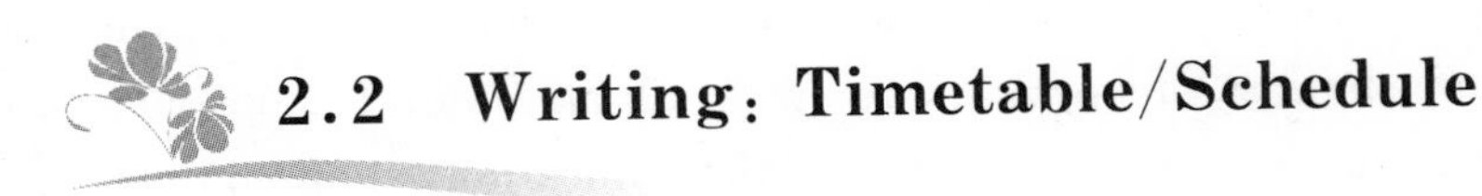

2.2 Writing: Timetable/Schedule

Timetable/Schedule

Introduction to Timetable: The timetable and schedule are used to arrange the schedule and travel itinerary of large-scale meetings and other activities, which are generally in written form. The schedule also includes bus, train, subway and flight schedules.

How to make a Schedule: At the beginning of the schedule, you should clearly indicate the arrangement plan of what activities it is in eye-catching font. In each item, the specific time shall be indicated first, and then the activity content, place or person in charge shall be written. When arranging, it can be written once in chronological order. Instead of using complete sentences, use noun phrases or participles for brevity.

Sample: Schedule for Professor David Smith

Schedule for Professor David Smith	
	February 8, 2022
11:00 noon	Pick-up at Rujia Hotel by Wang Hong
2:00 p.m.	Meeting at Conference Room of Building A (2th floor)
4:00～5:30 p.m.	Campus Tour
6:30 p.m.	Dinner at Beihai Restaurant
	February 9, 2022
8:00～11:00 a.m.	Lecture in Multimedia Room of Building A (1st floor)
11:30 noon	Lunch
1:00～5:30 p.m.	Tour of Hefei City
6:30 p.m.	Back to Rujia Hotel for Dinner
	February 10, 2022
9:00 a.m.	Check-out at Rujia Hotel
9:30 p.m.	To Xinqiao Airport (Terminal 3)

2.3 Careers in IT

Software testing engineer

Software testing engineer refers to the special staff who understand the functional requirements of the product, test it, check whether the software has defects (bugs), test whether the software has the performance of robustness, safety and operability, and write the corresponding test specifications and test cases. In short, software test engineers play the role of "quality management" in a software enterprise, find software problems in time and urge correction in time to ensure the normal operation of products. They are divided into three categories according to their levels and positions.

Senior software testing engineer, proficient in software testing and development technology, well aware of the industry of the tested software, and able to analyze and evaluate the possible problems.

Intermediate software test engineer, write software test plan and test documents, work with the project team to formulate the work plan of software test stage, and be able to reasonably use test tools to complete test tasks in project operation.

Junior software test engineer, whose job is usually to test the function of the product according to the software test scheme and process, and check whether the product has defects.

2.4 Words and phrases

allocate
alternative
application
bash
background
Basic Input/Output System (BIOS)
boot
boot loader
boot sector
by default

click and drag
compatibility
compress
configure
convention
coordination
database management
de-allocate
desktop background
device driver
dialog box
directory
display
document
error detecting
filesystem image
gallery
grainy
Grand Unified Boot loader (GRUB)
icons
initialize
install
interact with
kernel
Master Boot Records (MBR)
menu
motherboard
mount
navigation
performance
peripheral devices
personalize
pixelated
portion
Power On Self-Test (POST)
process management
profession
RAM
removable media
request
response
ribbon GUI
right-clicking
root filesystem
software system
specific task
splash screen
startup
stretched
system software
text-mode
toolbar
Unified Extensible Firmware Interface (UEFI)
utility
wallpaper
web-based application

2.5 Exercises

Ⅰ. Matching

Match each numbered item with the most closely related lettered item. Write your answers in the spaces provided.

a. application ________(1) one of a number of things from which only one can be chosen

b. menu ________(2) the ability of machines, especially computers, and computer programs to be used together

c. compatibility ________(3) the programs, used to operate a computer

d. alternative ________(4) computer software that performs a task or set of tasks, such as word processing or drawing

e. directory ________(5) the main board of a computer, containing all the circuits

f. icon ________(6) the background pattern or picture that you choose to have on your computer screen

g. motherboard ________(7) a list of possible choices that are shown on a computer screen of a program

h. software ________(8) a repository where all files are kept on computer

i. peripheral ________(9) a picture on a computer screen representing a particular computer function

j. wallpaper ________(10) any external device attached to a computer to enhance operation

Ⅱ. Written practice

Please make a schedule of activities for tomorrow according to your class schedule for tomorrow. It contains at least the following three-field information.

time interval	classroom	activities

Ⅲ. Open-ended

On a separate sheet of paper, respond to each question or statement.

(1) What are word processors? What are they used for?

(2) What are spreadsheets? What are they used for?

(3) What are presentation graphics programs? What are they used for?

(4) What are database management systems? What are they used for?

(5) Explain the difference between general-purpose and specialized applications. Also discuss the common features of application programs, including those with traditional and ribbon graphical user interfaces.

(6) Discuss general-purpose applications, including word processors, spreadsheets, database management systems, and presentation graphics.

(7) Discuss specialized applications, including graphics programs, video game design software, web authoring programs, and other professional specialized applications.

(8) Describe mobile Apps, including popular Apps and App stores.

(9) Describe software suites, including office suites, cloud suites, specialized suites, and utility suites.

Chapter 3 Data and database

Learning objectives

After you have read this chapter, you should be able to:

☆ Explain data and DIKW Pyramid model.
☆ Explain data in computer, including binary, units and numbers in computer.
☆ Describe image in computer, including digital image, pixel and color depth.
☆ Evaluate units of binary, including bit, byte, MB, KB, GB and others.
☆ Describe the relationship between data and database.
☆ Identify structured data and unstructured data.
☆ Define DBMS and RDBMS.
☆ Describe the NoSQL and its advantages.

3.1 Reading

Passage 1: The brief of data

Ⅰ. Data, information, knowledge and wisdom

In the conceptual world, the relationship among data, information, knowledge and wisdom can be described by the hierarchical pyramid model — DIKW Pyramid. Data is physical symbol. Only when data is recognized by human beings can it becomes information and so forth (See Figure 3.1).

In the DIKW Pyramid model (See Figure 3.2) each building block is a step towards a higher level — first comes data, then is information, next is knowledge and finally comes wisdom. Each step answers different questions about the initial data and adds value to it. The more we enrich our data with meaning and context, the more knowledge and insights we get out of it so we can take better, informed and data-based decisions.

Figure 3.1 Evolution of oracle bone inscriptions

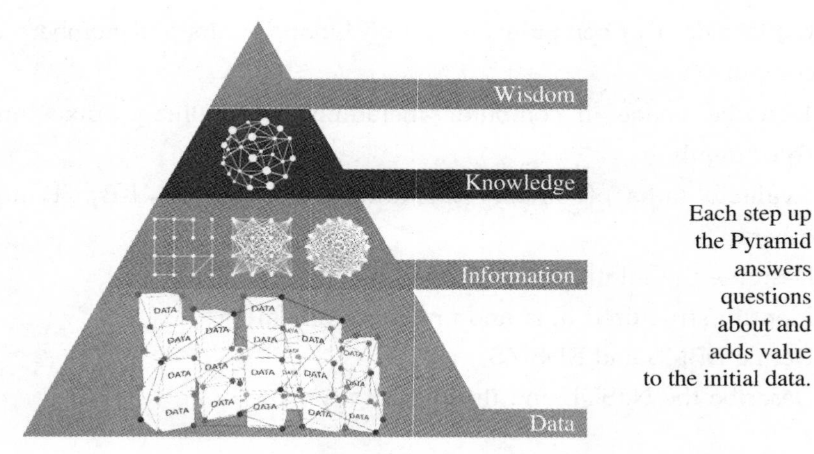

Figure 3.2 DIKW Pyramid model

Data is a collection of facts in a raw or unorganized form such as numbers or characters. However, without context, data can mean little (See Figure 3.3). For example, 12012022 is just a sequence of numbers without apparent importance. But if we view it in the context of "this is a date", we can easily recognize 12th of January, 2022.

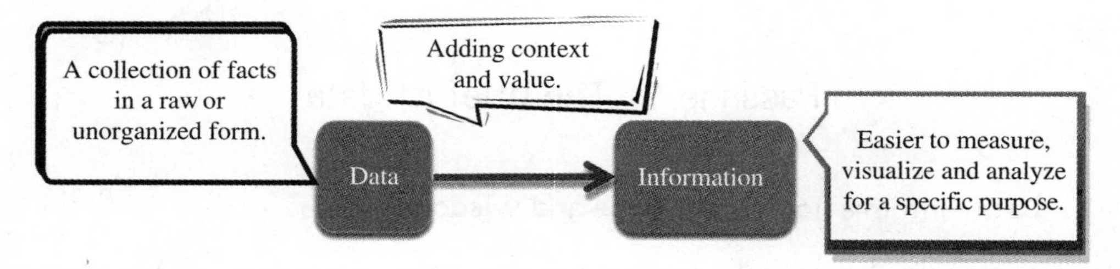

Figure 3.3 Data scale up information

By adding context and value to the numbers, they now have more meaning. In this way, we have transformed the raw sequence of numbers into information.

"How" is the information, derived from the collected data, relevant to our goals? "How" are the pieces of this information connected to other pieces to add more meaning and value? And, maybe most importantly, "How" can we apply the information to achieve our goal?

When we don't just view information as a description of collected facts, but also understand how to apply it to achieve our goals, we turn it into knowledge (See Figure 3.4).

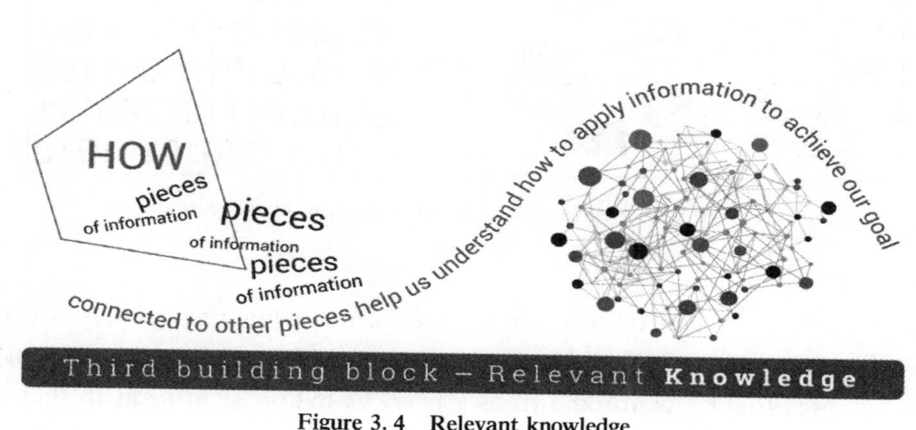

Figure 3.4 Relevant knowledge

This knowledge is often the edge that enterprises have over their competitors. As we uncover relationships that are not explicitly stated as information, we get deeper insights that take us higher up the DIKW Pyramid. But only when we use the knowledge and insights gained from the information to take proactive decisions, we can say that we have reached the final — "wisdom" — step of the Knowledge Pyramid (See Figure 3.5).

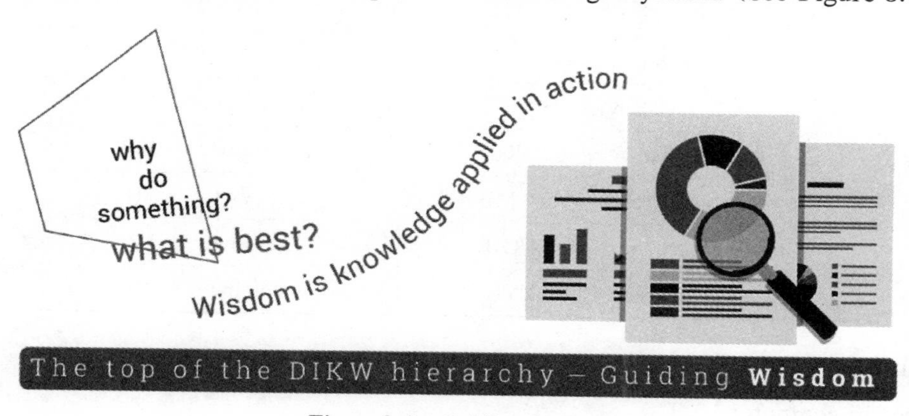

Figure 3.5 Guiding wisdom

Ⅱ. Data in computer

All data in a computer is represented in binary, whether it is numbers, text, images or sound. The computer software processes the data according to its content (See Figure 3.6).

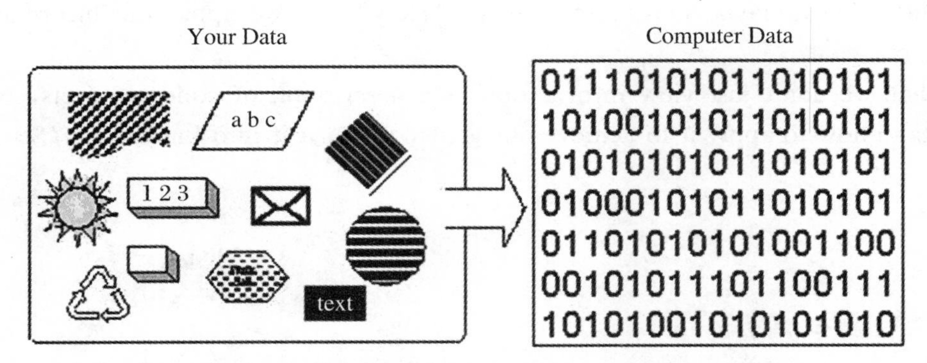

Figure 3.6 Representing real-world data in the computer

1. Units

In a computer, all data is stored in binary form. A binary digit has two possible states — 1 and 0. A binary digit is known as a bit. A bit is the smallest unit of data a computer can use. Eight bits are known as a byte. A byte is significant in that a single character can be represented in binary in eight bits — one byte.

Binary values are used to represent many kinds of data, namely numbers, text, images or sound. To be able to reference large numbers of 0s and 1s, the binary unit system is used (See Table 3.1):

Table 3.1 The binary unit system

Size	Unit
8 bits	1 byte
1,024 bytes	1 kilobyte — KB
1,024 kilobytes — 1,024 KB	1 megabyte — MB
1,024 megabytes — 1,024 MB	1 gigabyte — GB
1,024 gigabytes — 1,024 GB	1 terabyte — TB
1,024 terabytes — 1,024 TB	1 petabyte — PB

Four bits or half a byte is known as a nibble.

2. Images in computer

Computers can only recognize binary. All data must be converted into binary in order for a computer to process it. Images are no exception. The digital images are made up of pixels. Each pixel is represented by a binary number.

How computers process and represent images? Consider a simple black-and-white image. If 0 is black and 1 is white, then a simple black-and-white picture can be created using binary.

To create the picture, a grid can be laid out and displayed on a screen. The squares on the screen, known as pixels, are colored (0 — black and 1 — white). The following image is a representation of how this picture would be displayed on a screen (See Figure 3.7).

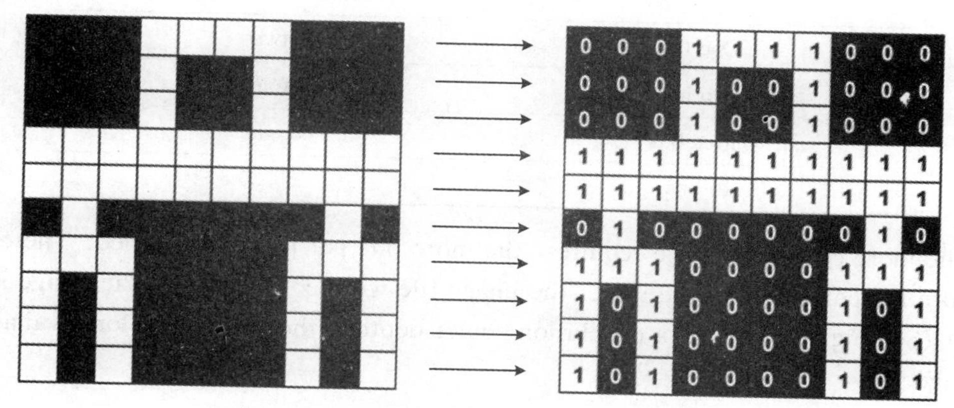

Figure 3.7 Black-and-white image in computer

But many images need to use colors. To add color, more bits are required for each pixel. The number of bits determines the range of colors. This is known as an image's color depth.

For example, using a color depth of two, i. e, two bits per pixel, allows for four possible colors, such as: 00 (represents black), 01 (represents dark grey), 10 (represents light grey) and 11 (white). The image will be represented in computer as Figure 3.8.

00	00	00	11	11	11	11	00	00	00
00	00	00	11	00	00	11	00	00	00
00	00	00	11	00	00	11	00	00	00
01	01	01	01	01	01	01	01	01	01
01	01	01	01	01	01	01	01	01	01
00	10	00	00	00	00	00	00	10	00
10	10	10	00	00	00	00	10	10	10
10	00	10	00	00	00	00	10	00	10
10	00	10	00	00	00	00	10	00	10
10	00	10	00	00	00	00	10	00	10

Figure 3.8 Four-color image represented in computer

Each extra bit doubles the range of colors that are available (See Table 3.2).

Table 3.2　Bits per pixel of image

Bits per pixel	Possible colors counter
One bit	two
Two bits	four
Three bits	eight
16 bits	65536

The more colors an image requires, the more bits per pixel are needed. Therefore, the more the color depth, the larger the image file will be. Of course, the image is also clearer (See Figure 3.9). In a conclusion, color depth is the range of colors available in an image.

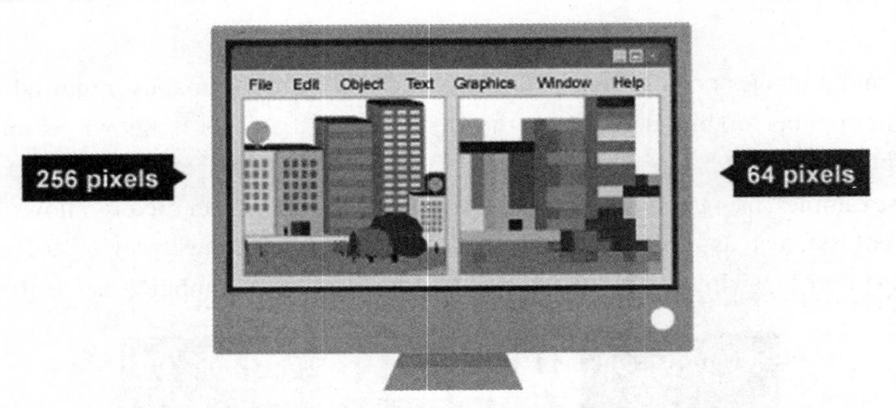

Figure 3.9　Clarity of different pixel images

Passage 2: Why use a database

Database are used to store data in a structured or organized format. The main purpose of using database is to share data among a variety of application systems (See Figure 3.10). For example, database are used in many different ways, from a supermarket tracking stock in a store to the contacts list in a mobile phone. Other kinds of data stores can also be used, such as files on the file system or large hash tables in memory but data fetching and writing would not be so fast and easy with those type of systems.

At present, the models for storing data mainly include relational database and non-relational database.

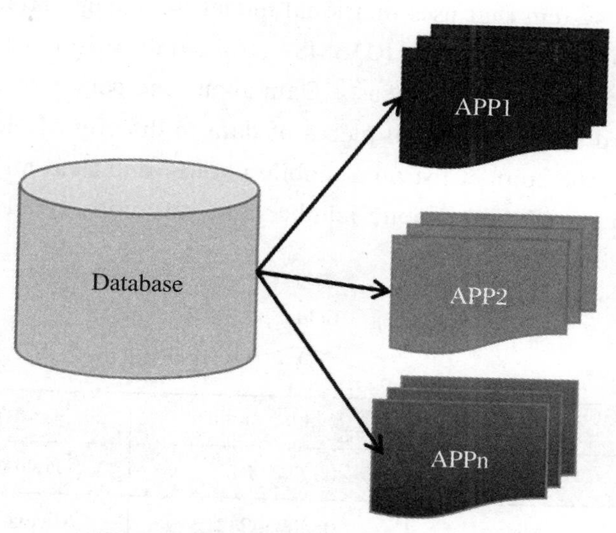

Figure 3.10 Database sharing the data

Ⅰ. Structured data Unstructured data

Data is said to be unstructured if it isn't recorded in a predefined, organized way. For example, information written down in a notebook or into a word-processed file with no organization would be unstructured data.

The opposite of unstructured is structured. Structured data would be stored using predefined headings to organize the data into a particular format (See Figure 3.11).

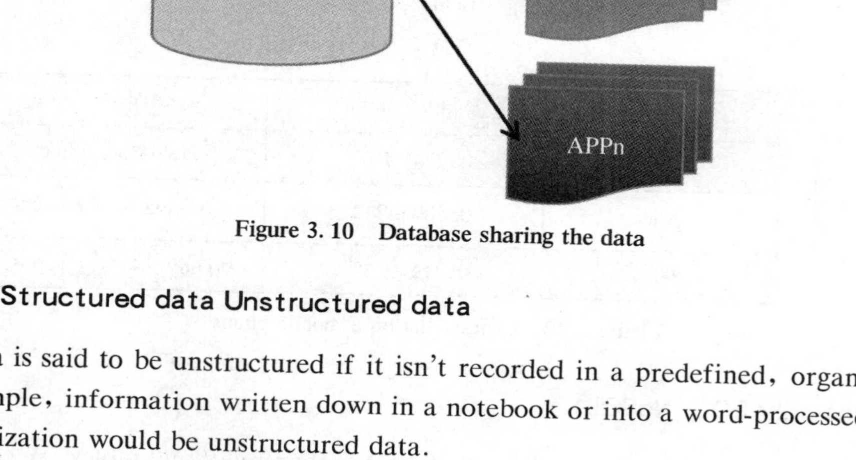

Figure 3.11 Unstructured and structured data

Ⅱ. Relational Database Management System (RDBMS)

2-dimensional table is a common structure for storing data. The structure is called a relation model. The system that uses relational model to manage database is called Relational Database Management System (RDBMS). If a database is managed using RDBMS, the database is called a relational database. Data about one person or thing will be stored in a record. A record stores individual pieces of data in different fields.

For instance, in the contact list on a mobile phone, one field might store a person's name, another field stores their mobile number and yet another field stores their email address (See Figure 3. 12).

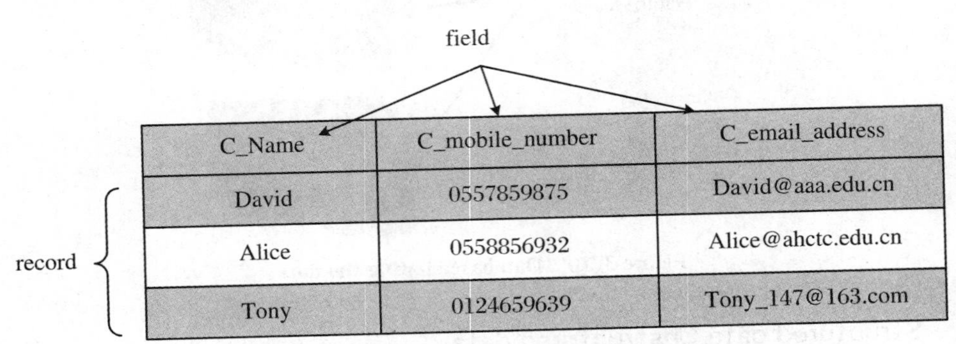

Figure 3. 12 Contact list on a mobile phone

Ⅲ. Not only SQL (NoSQL)

NoSQL databases store data in documents rather than relational tables. Accordingly, we classify them as "not only SQL" and subdivide them by a variety of flexible data models. Types of NoSQL databases include pure document databases, key-value stores, wide-column databases, and graph databases. NoSQL databases are built from the ground up to store and process vast amounts of data at scale and support a growing number of modern businesses.

Customer experience has quickly become the most important competitive differentiator and ushered the business world into an era of monumental change. As part of this revolution, enterprises are interacting digitally — not only with their customers, but also with their employees, partners, vendors, and even their products — at an unprecedented scale. This interaction is powered by the internet and other 21st century technologies — and at the heart of the revolution are a company's cloud, mobile, social media, big data, and IoT applications.

How are these applications different from legacy enterprise applications like ERP, HR, and financial accounting? Today's web, mobile, and IoT applications share one or more (if not all) of the following characteristics. They need to:

(1) Support large numbers of concurrent users (tens of thousands, perhaps millions).

(2) Deliver highly responsive experiences to a globally distributed base of users.

(3) Be always available — no downtime.

(4) Handle unstructured data.

(5) Rapidly adapt to changing requirements with frequent updates and new features.

Building and running these massively interactive applications has created a new set of technology requirements. The new enterprise technology architecture needs to be far more agile than ever before, and requires an approach to real-time data management that can accommodate unprecedented levels of scale, speed, and data variability. Relational databases are unable to meet these new requirements, and enterprises are therefore turning to NoSQL database technology.

The five critical differences between SQL and NoSQL are:

(1) SQL databases are relational, NoSQL databases are non-relational.

(2) SQL databases use structured query language and have a predefined schema. NoSQL databases have dynamic schemas for unstructured data.

(3) SQL databases are vertically scalable, while NoSQL databases are horizontally scalable (See Figure 3.13).

Figure 3.13 SQL VS NoSQL

(4) SQL databases are table-based, while NoSQL databases are document, key-value, graph, or wide-column stores.

(5) SQL databases are better for multi-row transactions, while NoSQL is better for unstructured data like documents or JSON.

So, what are NoSQL databases and why do they matter now? As enterprises shift to the digital economy — enabled by cloud, mobile, social media, and big data technologies — developers and operations teams have to build and maintain web, mobile, and IoT applications faster and faster, and at greater scale. Flexible, high-performance NoSQL is increasingly the preferred database technology to power today's web, mobile, and IoT applications.

Hundreds of Global 2000+ enterprises, along with tens of thousands smaller businesses and startups, have adopted NoSQL. For many, the use of NoSQL started with a cache, proof of concept, or a small application, then expanded to targeted mission-critical applications, and is now the foundation for all application development.

With NoSQL, enterprises are better able to both develop with agility and operate at any scale — and deliver the performance and availability required to meet the demands of digital economy businesses.

3.2 Writing: How to write job applications

Job application

Employers may receive hundreds of applications for a job, so it's vital to make sure that the letter or email you send with your resume creates the right impression. It's your opportunity to say why you want the job, and to present yourself as a candidate for the post in a way that impresses a prospective employer and makes you stand out as a prospective employee.

Ⅰ. Preparation

Before you start:

Read the advert closely so that you can tailor your application to the requirements of the job.

Research the organization: this will show prospective employers that you really are interested in them.

Ⅱ. Composing the letter or email

General points:

- Keep it brief. You don't need to give a lot of detail. What you are aiming for is a clear and concise explanation of your suitability for the job.
- Begin your letter or email with "Dear Mr./Mrs./Ms. ××××" if you know the person's name, or "Dear Sir or Madam" if you don't know their name.
- Avoid inappropriate language such as slang or technical jargon.
- Use brief, informative sentences and short paragraphs.
- Check your spelling, grammar, and punctuation carefully. Some employers routinely discard job applications that contain such mistakes.

Ⅲ. Structure

The usual order of a job application letter or email is:

The position applied for: Give the title of the job as a heading, or refer to it in the first sentence of your letter, using the reference code if there is one. This will ensure that your application goes directly to the right person in the organization. You should also mention where you saw the job advert or where you heard about the vacancy. If you heard about it through someone already working for the company, mention their name and position.

Your current situation: If you're working, briefly outline your current job. Pick up on the job requirements outlined in the advert and focus on any of your current skills or responsibilities that correspond to those requested. For example, if the advert states that management skills are essential, then state briefly what management experience you have. If you're still studying, focus on the relevant aspects or modules of your course.

Your reasons for wanting the job: Be clear and positive about why you want the job. You might feel that you are ready for greater challenges, more responsibility, or a change of direction, for example. Outline the qualities and skills that you believe you can bring to the job or organization.

Closing paragraph: In the final paragraph you could say when you'd be available to start work, or suggest that the company keep your resume on file if they decide you're not suitable for the current job.

Signature: If you are sending a letter rather than an email, always remember to sign it and to type your name underneath your signature.

Sample: Application of database administrator

Please write a letter to apply for the position of database administrator in HUAWEI.

Dear Hiring Manager,

I am delighted for the opportunity to apply for the position of database administrator in HUAWEI. I have enjoyed my time working as a database professional during my career, and I have met and solved many informational and data-related challenges in the process. I am always interested in acquiring better ways to improve database efficiency and performance, and I look forward to learning more about the goals that HUAWEI wants to achieve in the feild of safe and secure data management.

As a senior database administrator, I have maintained multiple database systems for my employers. While at former company I headed the team of database professionals whose jobs were to monitor, design and streamline all the repositories of data. This data was highly important and contained many personal details about our customers' financial information. I ensured that all safety protocols were followed and that best practices were implemented at all times.

Thank you for your consideration of my application. I am interested in hearing more details about the database administrator position and learning more about HUAWEI. I am always ready to share my knowledge with others, and I love learning new things from my peers, especially in a fast-paced environment such as HUAWEI.

I look forward to hearing from you in the near future.

Sincerely,

David

3.3 Careers in IT

Data analysts

The data analyst serves as a gatekeeper for an organization's data so stakeholders can understand data and use it to make strategic business decisions. It is a technical role that requires an undergraduate degree or master's degree in analytics, computer modeling, science, or math.

A master of professional studies in analytics prepares students for a career as a data analyst by covering the concepts of probability theory, statistical modeling, data visualization, predictive analytics, and risk management in the context of a business environment. In addition, a master's degree in analytics equips students with the programming languages, database languages, and software programs that are vital to the day-to-day work of a data analyst.

Database administrators

Database administrators use database management software to determine the most efficient ways to organize and access a company's data. Additionally, database administrators are typically responsible for maintaining database security and backing up the system. Database administration is a fast-growing industry and substantial job growth is expected.

Database administrator positions normally require a bachelor's degree in computer science or information systems and technical experience. Internships and prior experience with the latest technology are a considerable advantage for those seeking jobs in this industry. It is possible to transfer skills learned in one industry, such as finance, to a new career in database administration. In order to accomplish this objective, many people seek additional training in computer science.

3.4 Words and phrases

back up
big data
binary
byte
cloud
collection
color depth
context
dark grey
data analysts
Database Administrator (DBA)
data-based decision
dynamic
enrich
field
gigabyte
graph databases
hierarchical pyramid model
horizontal
information
initial data
interactive application
IoT applications
key-value stores
kilobyte
megabyte
nibble
non-relational database
NoSQL
petabyte
physical symbol
pixel
pure document databases
raw sequence

RDBMS
record
relational database
relationship
social media
SQL
structured data
subdivide
terabyte
unstructured data
value
vertical
wide-column databases

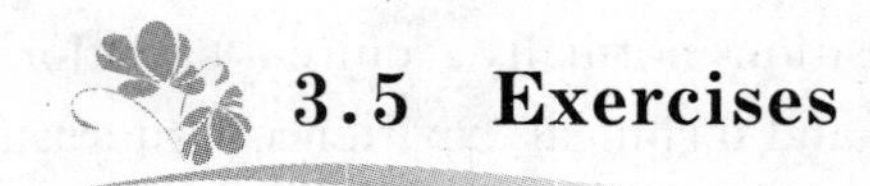

3.5 Exercises

Ⅰ. Matching

Match each numbered item with the most closely related lettered item. Write your answers in the spaces provided.

a. attributes ________(1) view that focuses on the actual format and location of the data

b. field ________(2) type of processing in which data is collected over several hours, days, or even weeks and then processed all at once

c. batch ________(3) another name for a data dictionary

d. physical ________(4) type of database structure where the data elements are stored in different tables

e. redundancy ________(5) object-oriented databases organize data by classes, objects, methods

f. hierarchical ________(6) group of related characters

g. relational ________(7) a data problem that often occurs when individual departments create and maintain their own data

h. speed ________(8) type of database structure where fields or records are structured in nodes that are connected like the branches of an upside-down tree

i. schema ________(9) type of database that uses communication networks to link data stored in different locations

j. redundancy ________(10) two of the most significant advantages of multi-dimensional databases are conceptualization and processing

Ⅱ. Written practice

HUAWEI is recruiting a programmer for its business and you're interested in this position. Please write a letter of application to apply for the position.

Ⅲ. Open-ended

On a separate sheet of paper, respond to each question or statement.

(1) Explain the data, information, knowledge and wisdom.
(2) Describe the relationship between data and database.
(3) What is a NoSQL?

Chapter 4 Communications and networks

Learning objectives

After you have read this chapter, you should be able to:

☆ Explain communication and computer networks.
☆ Describe computer network resources.
☆ Compare different networks, including LAN, WAN and Internet.
☆ Describe the relationship between communication system and network.
☆ List the connection devices.
☆ Explain Internet of things and Internet.

Communication networks are the backbone of nearly every aspect of modern digital life (See Figure 4.1). In the future, telepresence — the ability to fully experience the reality of a different place without actually being there — will be commonplace. For example, you will routinely perform housework located halfway around the world. Maybe you are undergoing.

Figure 4.1 Digital life

Computer network is one of the most far-reaching scientific and technological inventions that have brought human civilization. At present, there are many network-based applications and services in your life.

As the power and flexibility of our communication systems have expanded, the sophistication of the networks that support these systems has become increasingly critical and complex. The network technologies that handle our cellular, business, and Internet communications come in many different forms. Satellites, broadcast towers, 5G, even buried cables and fiber optics carry our telephone messages, e-mail, and text messages. These different networks must be able to efficiently and effectively integrate with one another.

To efficiently and effectively use computers, you need to understand the concept of connectivity, wireless networking, and the elements that make up network and communications systems. Additionally, you need to understand the basics of connection devices, data transmission, network types and network architectures.

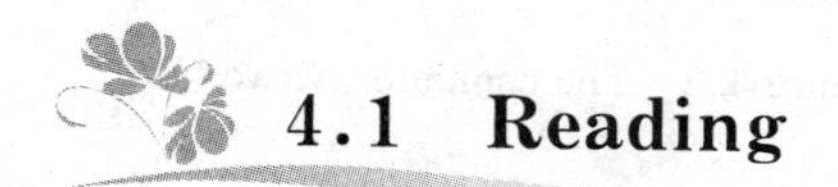

4.1 Reading

Passage 1: The computer network

Ⅰ. Brief of network

1. Definition

The computer network is a communication system that connects two or more computers so that they can exchange information and share resources. It can be set up in different arrangements to suit user's needs (See Figure 4.2). An internet itself can be considered the computer network.

In computer networks, the computing devices exchange the data with each other by using the connections between nodes. The data links are established over the communication resources, including cable media such as wires or optic cables or wireless media such as Wi-Fi.

The computer network is also a digital telecommunication network which allows nodes to share the resources. It is the group of computer systems and other computing hardware devices. Those devices are linked together through communication channels to facilitate communication and resource sharing among a wide range of users.

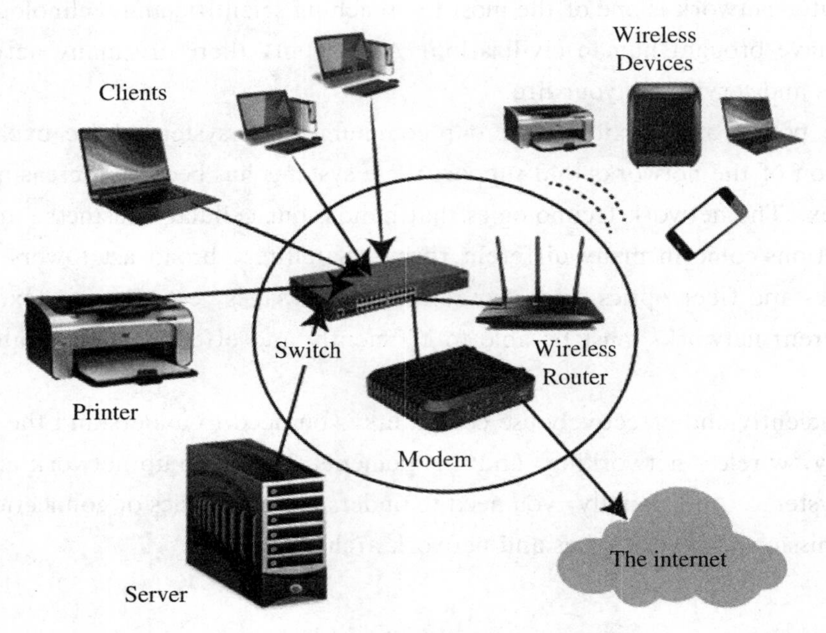

Figure 4. 2 The computer network

2. Classification of network resources

Everything in the network is called network resources. There are classified two categories, including computing resources and communication resources. The computing resources include: clients, server, printer, etc. The computing nodes of a computer network may include personal computers, servers, other specific or general-purpose hosts. They are identified by network addresses, and may have host-names. A communications resource is a physical or logical device that provides a single bidirectional, asynchronous data stream. Serial ports, parallel ports, fax machines, and modems are examples of communications resources. For each communications resource, there is a service provider, consisting of a library or driver, that enables applications to access the resource.

Networks are used to facilitate communication via email, video conferencing, instant messages, etc. It enables files sharing across the Network. These are also used to make information easier to access and maintain among the network users.

Ⅱ. Types of networks

The Network is also used to enable the multiple users to share a single hardware device like the printer or scanner. It allows for sharing of software or operating programs on the remote system.

There are multiple devices or medium in a computer network, which helps in the communication between two different devices that are called as network devices, for example, router, switch, bridge, hub, gateway, modem, repeater. The Network is the interconnection of multiple devices and termed as hosts connected by using multiple paths to send and receive the data or media. The computer network is mainly of four types, which are given below (See Figure 4. 3):

- MAN (Metropolitan Area Network)
- PAN (Personal Area Network)
- LAN (Local Area Network)
- WAN (Wide Area Network)

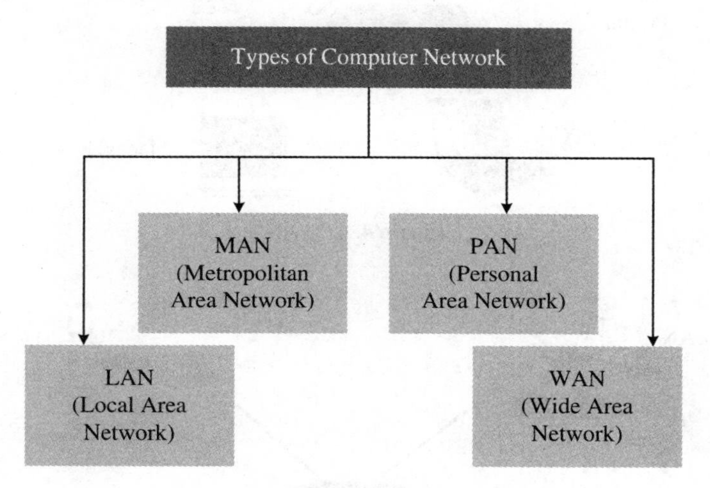

Figure 4. 3 Types of computer network

There are several types of protocols which are used in networking. The protocol is a set of rules or algorithms which defines the way how two entities can communicate across the Network, and different protocol is defined at every layer of the OSI Model.

1. LAN (Local Area Network)

The Local Area Network is a computer network which interconnects the computer within the limited area such as a residence, school laboratory, university campus or office building. LAN spans a relatively small area. In a wireless LAN, the users have unrestricted movement within the coverage area (See Figure 4. 4).

The wireless networks become popular in the residences and small businesses just because of their easy installation. The network topology describes the layout of interconnections between devices and network segments. The wide varieties of LAN topologies are used at the data link layer and physical layer.

2. MAN (Metropolitan Area Network)

The Metropolitan Area Network is a computer network which is similar to the Local Area Network but spans an entire city or campus (See Figure 4. 5).

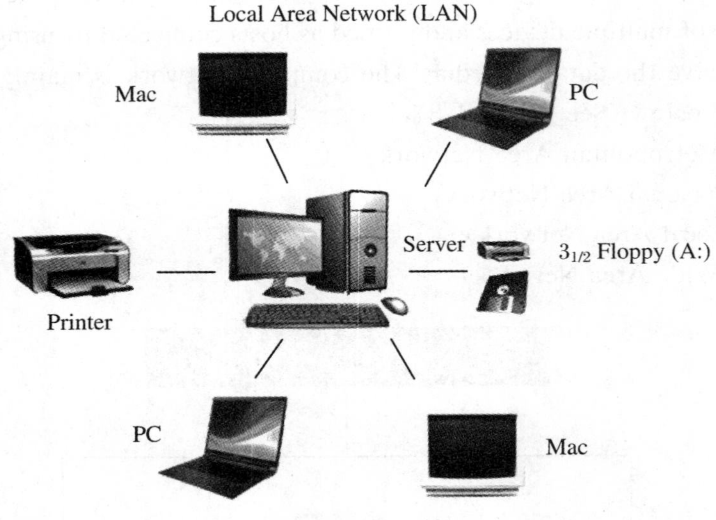

Figure 4. 4 LAN

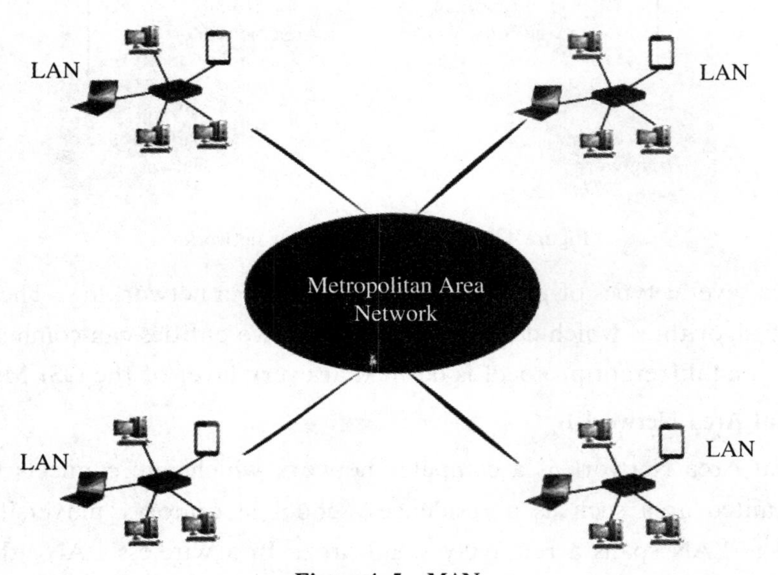

Figure 4. 5 MAN

In Metropolitan Area Network, we interconnect the user with computer resources in a whole geographic area or region. This Network is larger or broader than the Local Area Network.

3. PAN (Personal Area Network)

The Personal Area Network is used for interconnecting the devices which are centered on a person's workspace (See Figure 4. 6). This Network provides data transmission among devices such as computers, smartphones, tablets, and personal digital assistants.

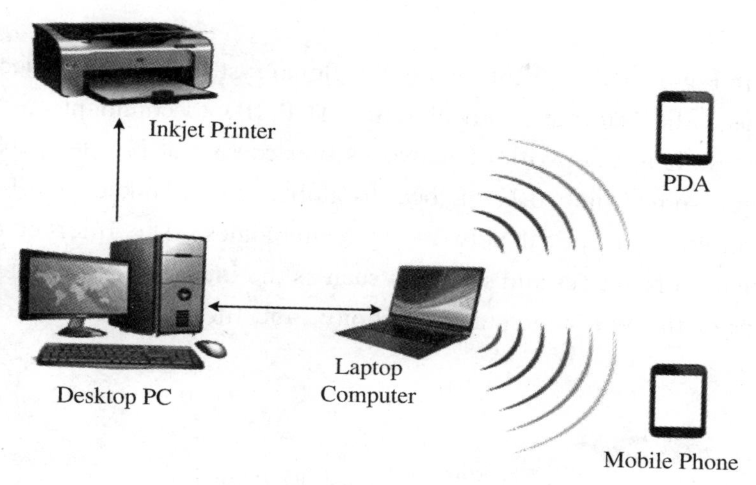

Figure 4.6 PAN

4. WAN (Wide Area Network)

The Wide Area Network is the telecommunication network which extends over the large geographical area for the primary purpose of computer networking (See Figure 4.7). These networks are often established with leased telecommunication circuits. The WAN connects different smaller networks, including the Local Area Network.

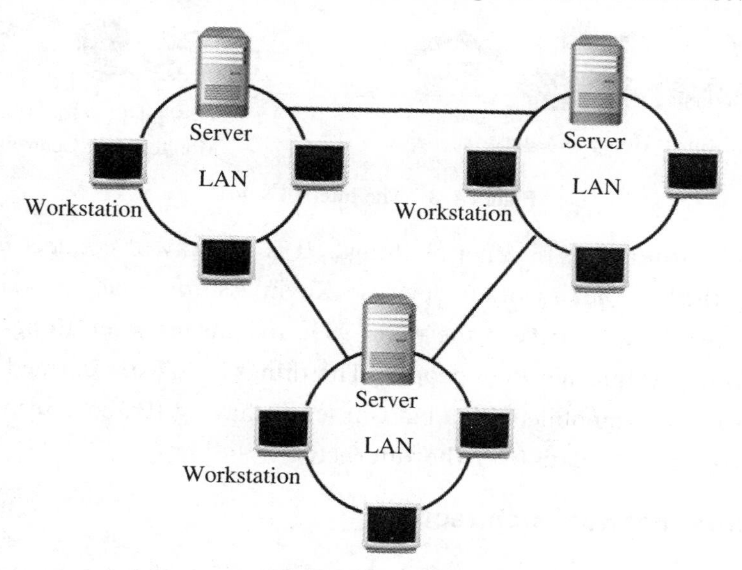

Figure 4.7 WAN

The computers connected to the Wide Area Network are often connected through public networks, such as telephone systems. This Network can connect through the leased lines or satellites.

5. Internet and IoT

As shown in Figure 4.8, the Internet is the global system of interconnected computer networks that uses the Internet protocol suite (TCP/IP) to communicate between networks and devices. It is a network of networks that consists of private, public, academic, business, and government networks of local to global scope, linked by a broad array of electronic, wireless, and optical networking technologies. The Internet carries a vast range of information resources and services, such as the inter-linked hypertext documents and applications of the WWW, email, telephony, and file sharing.

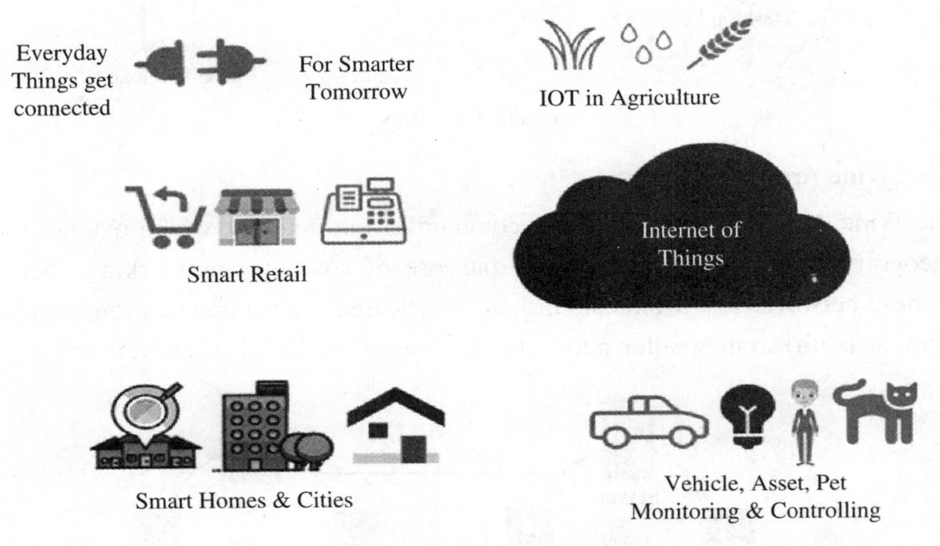

Figure 4.8 The internet & IoT

IoT is also known as the Internet of things. It is the way of connecting the physical objects through the internet to other devices. Kevin Ashton gave the term internet of things, and he mentioned it first in the year 1999. IoT means when things or objects are more connected to the Internet than people. The things in IoT are defined as objects that can be the person or automobile with a built-in sensor having IP addresses with the ability to collect and transfer the data over the Internet.

Ⅲ. Computer network architecture

Computer network architecture is a design in which all the computers are organized in a computer network. The architecture defines how computers must communicate with each other to obtain maximum benefits from a computer network, such as better response time, security, scalability, transfer data rate, connectivity, etc. There are two most popular computer network architectures used.

1. Peer-to-Peer Network

In a peer-to-peer Network, each computer acts as its client and server, i.e., it can perform both requests and responses. A Peer-to-Peer Network has no dedicated servers, but all computers act as a server for the data stored in them.

Peer-to-Peer Network is a network to which all computers are used the same resources and rights as other computers. Its network designed primarily for the small local area.

The following Figure depicts a Peer-to-Peer Network, with each computer network acting as both client and server (See Figure 4.9).

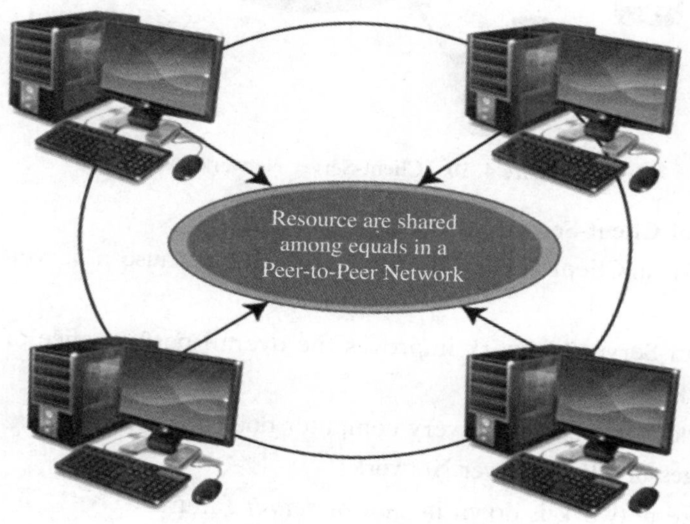

Figure 4.9 Peer-to-Peer Network Model

Advantages of Peer-to-Peer Network:

- Less costly: There are no dedicated servers, so the network is cost efficient.
- Simple setup & maintain: It is simple to set up and maintain.
- A network administrator is not required.

Disadvantages of Peer-to-Peer Network:

- Security is another problem on this network because malware can easily be transmitted through the network.
- It usually doesn't work well with more than "10" users.
- We can't access the shared data once the computer crashes or automatically turns off.

2. Client-Server Network

In a Client-Server Network, one central computer act as a hub that is known as a server, and all other computers are known as a client. A Client-Server Network has a dedicated server provider. All shared data is stored in the server, which is shared with the client computer when the client computer makes a request. The server is responsible for managing all data, such as files, directories, printers, etc. All clients connect via a server.

The following figure depicts a Client-Server Network (See Figure 4.10).

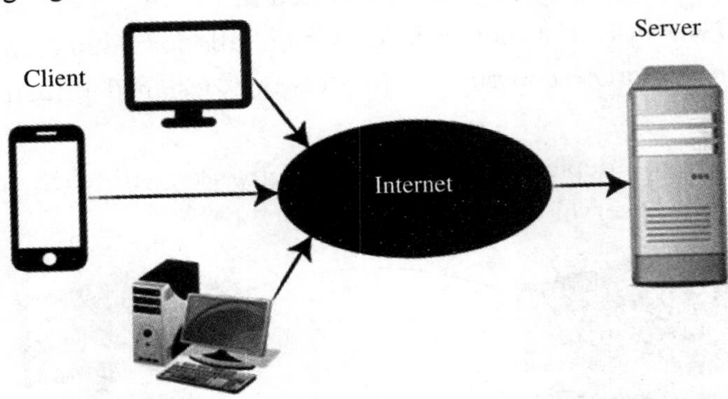

Figure 4.10 Client-Server Network Model

Advantages of Client-Server Network:

- Security in a Client-Server Network is better because a server manages shared resources.
- The Client-Server Network improves the overall performance of the system with the help of a dedicated server.
- Data backup is secure, as every computer does not need to manage the backup.

Disadvantages of Client-Server Network:

- The entire network is down in case of server failure.
- It difficult to set up and maintain.
- The Client-Server Network is costly, as it requires a large memory server.

The difference between Peer-to-Peer Network and Client-Server Network are described as following table (See Table 4.1).

Table 4.1 Differences between P2P and CS Network

Peer-to-Peer Network	Client-Server Network
It is used in small networks with less than "10" computers.	It usually used on both small and large networks.
In Peer-to-Peer Network, each computer is used to store its own data.	A centralized server is used to store data in a Client-Server network.
A Peer-to-Peer Network is cheaper than a Client-Server Network.	A Client-Server Network is more expensive than a Peer-to-Peer Network.
A Peer-to-Peer Network is less stable and secure than Client-Server Network.	A Client-Server Network is more stable and secure than Peer-To-Peer.
A Peer-to-Peer Network doesn't need a server.	A powerful computer that acts as a server.

Passage 2: How to set up a Wi-Fi network

The Internet is a really powerful tool. It gives us access to all kinds of information at a moment's notice — think email, Baidu search, and Wikipedia. So, there's something a little counter intuitive about only being able to use the Internet when you sit down at a desktop computer. What if you could use the Internet from anywhere in your home or office?

If you already have high-speed (broadband) Internet service at your house, it's pretty easy to create your own home wireless network. Commonly known as Wi-Fi, a wireless network (See Figure 4. 11) allows you to connect laptops, smartphones, and other mobile devices to your home Internet service without an Ethernet cable.

Figure 4. 11 Wireless network logic model

Follow the installation steps below to enjoy the wireless network services.

Step 1: Purchase a wireless router

To create your own Wi-Fi network, you'll need a wireless router. The device will broadcast the Wi-Fi signal from your Internet modem throughout your house. Your Internet Service Provider (ISP) may offer you a wireless router for a small monthly fee. If you've never set up a Wi-Fi network before, this may be the easiest option.

If you want to buy your own router, we'd recommend spending a little more time researching different options. We recommend choosing a wireless router based on broad-band size, room size, and number of connected devices.

Some Internet modems may already have a built-in wireless router, which means you won't need to purchase a separate one.

Step 2: Connect the cables

Once you've acquired a wireless router, you'll need to connect it to your existing Internet modem.

Connect an Ethernet cable from your modem to the wireless router (there is usually a short Ethernet cable included with your wireless router for this purpose). Plug in the power cable for the wireless router (See Figure 4.12). Wait at least 30 to 60 seconds, and make sure the lights on your router are working correctly.

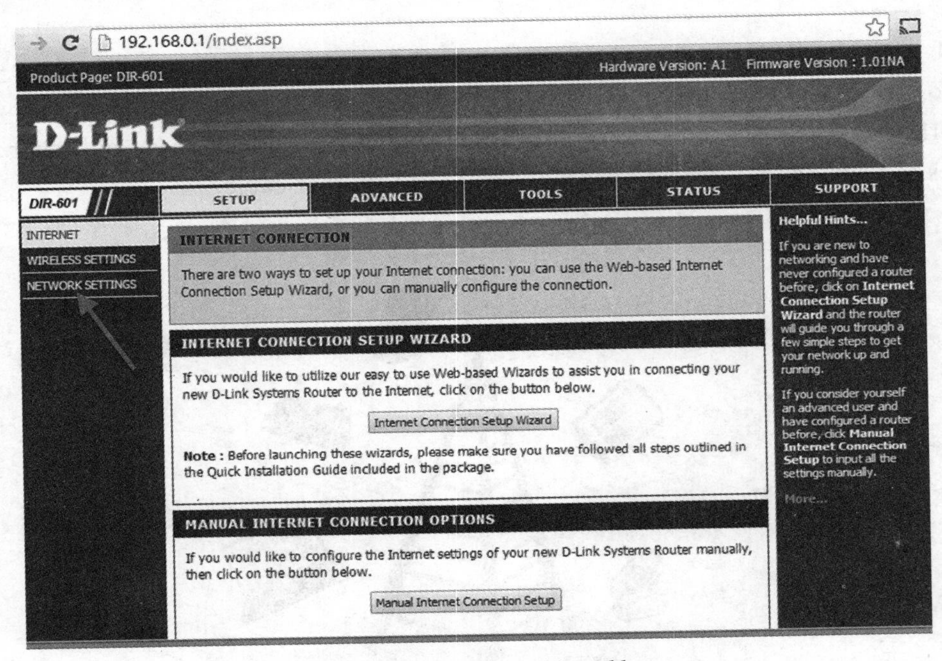

Figure 4.12 Connect the cables

Step 3: Configure your router

Next, you'll need to use your computer to configure your router's default settings. This includes setting a unique name and password for your wireless network (See Figure 4.13).

Using your web browser, enter the router's default IP address into the address bar, then press Enter. Your router's instructions should include this information, but some of the most common addresses include 192.168.0.1, 192.168.1.1, and 192.168.2.1.

The router's sign-in page will appear. Again, the exact sign-in details should be included with your router's instructions, but most routers use a standard user name and password combination, such as admin and password.

Your router's settings page will appear. Locate and select the network name setting, then enter a unique network name. Locate and select the network password setting, and choose an encryption option. There are several types of encryptions you can use, but we recommend WPA2, which is generally considered to be the most secure (See Figure 4.14).

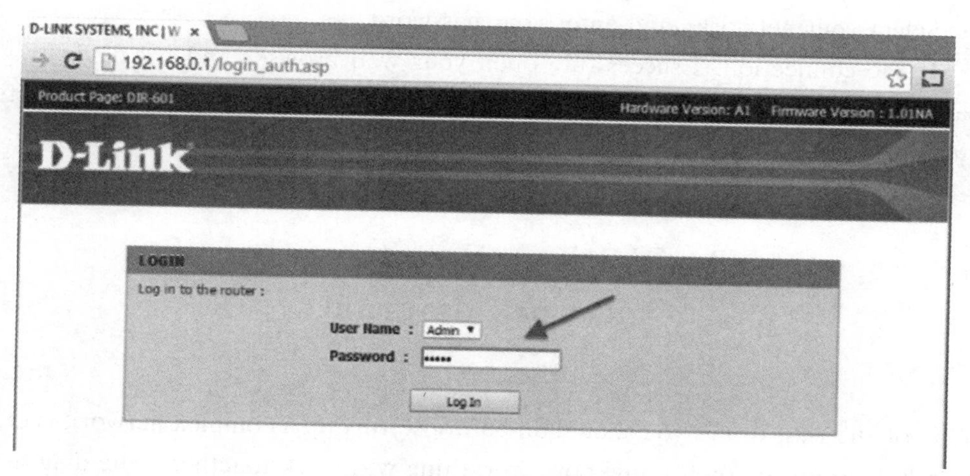

Figure 4.13 Login page

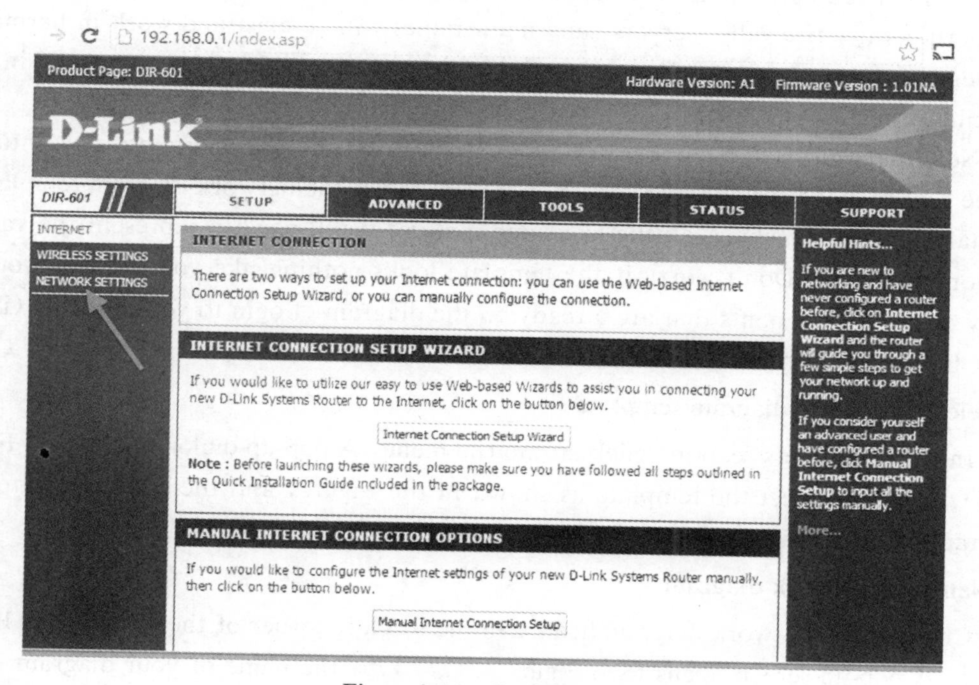

Figure 4.14 Configuration

Enter your desired password. Make sure to use a strong password to help ensure no one else can access your network. Locate and select the save button to save your settings.

Step 4: Connect testing

Now you're ready to connect to your Wi-Fi network and make sure it's working. The process for connecting to a Wi-Fi network will vary slightly depending on what type of computer or device you're using, but any system will require these basic steps:

- Locate your computer's network settings, and search for nearby Wi-Fi networks.
- Select your network, and enter your password.
- If the connection is successful, open your web browser and try navigating to a webpage like www.baidu.com. If the page loads, it means your Wi-Fi connection is working correctly.

4.2 Writing: How to create a network diagram

One of the first things you should do before setting up a complex network is creating a network diagram so you'll know how everything will work together. The diagram provides a visual representation of a network architecture. You can clearly see how things like peripherals, firewalls, servers, and mainframes will co-exist and work in harmony. Conversely, when a network doesn't work properly, this type of diagram can aid in pinpointing issues.

Now that you have a sketch of your network diagram, use diagrams.net's intuitive online tool to make a digital version. You'll want to use the network diagram template to get started since it includes all the icons and images, you'll need to represent the various network elements. Don't worry if the template looks nothing like your design, you can easily delete the elements that are already on the diagram. Login to your account (if you don't have one, sign up for a free draw) and follow the steps below.

1. Select a network diagram template

In the documents section, click on the file menu, A pop-up dialog opens (See Figure 4.15). You can select the template as shown in the Figure, and then click "create". A diagram will be created as showing in Figure 4.16.

2. Name the network diagram

Click on the network diagram header in the top left corner of the screen (See Figure 4.15). A pop-up screen opens (See Figure 4.17), type the name of your diagram in the text box and click Rename. The name of your network diagram appears in the top left corner of the screen.

3. Remove existing elements that you don't need on your diagram

A template is just a starting point, but if there are any elements on the network diagram template that you won't be using, remove them now. Click on the item and then right-click on the mouse. Menu options will appear on the screen, select Delete (See

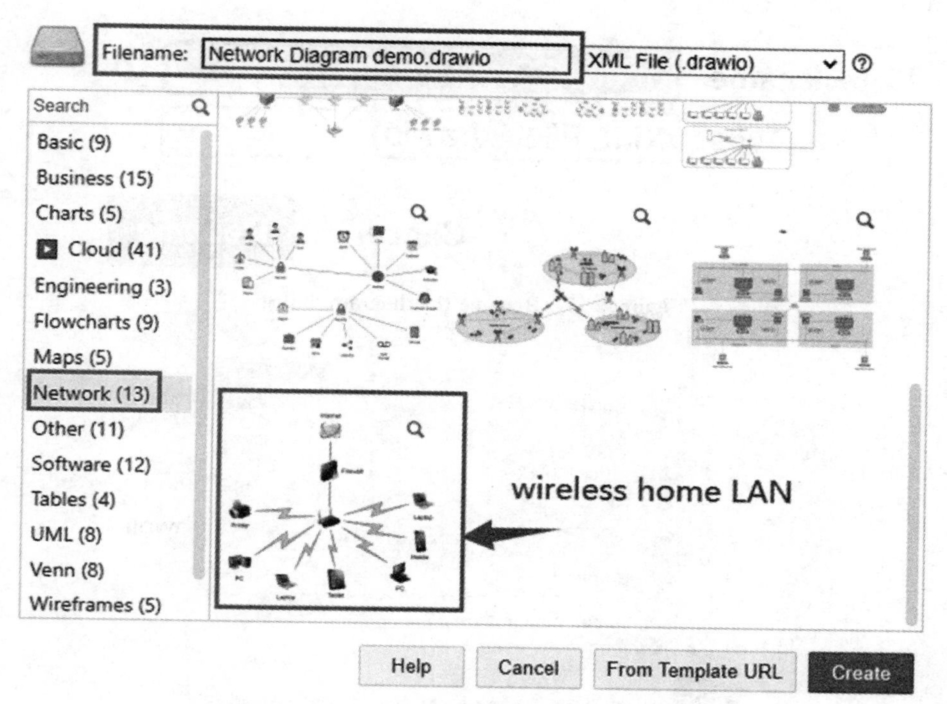

Figure 4.15 Select network diagram template

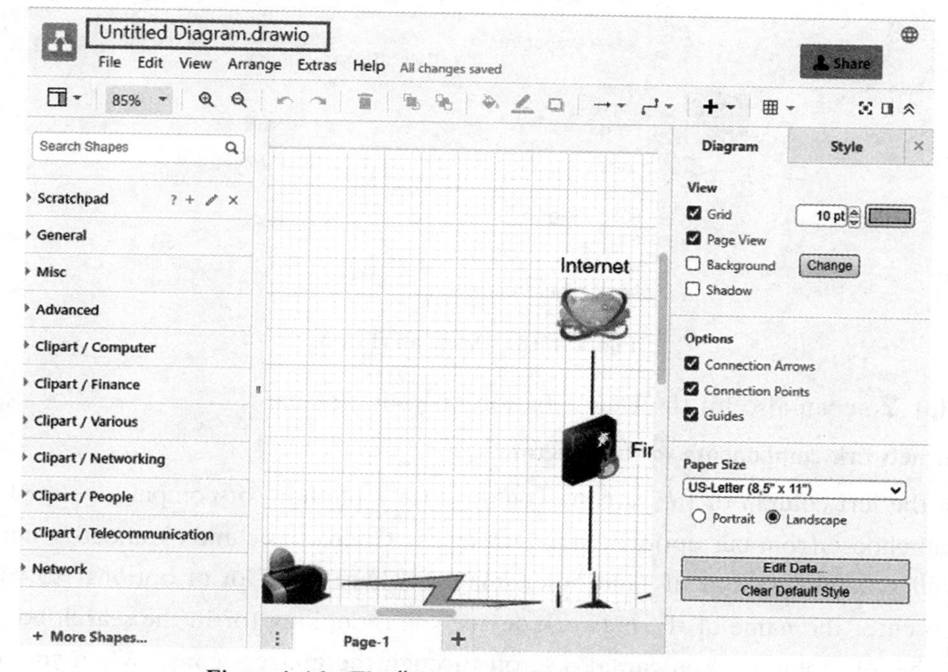

Figure 4.16 The "Unititled Diagram. drawio" Page

Figure 4.18).

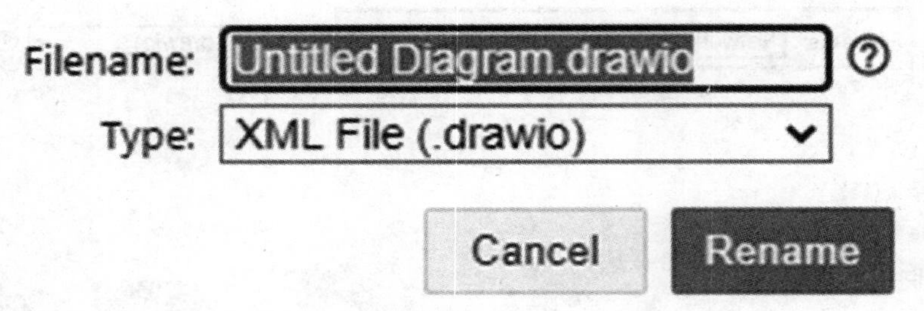

Figure 4.17 Rename the diagram dialog

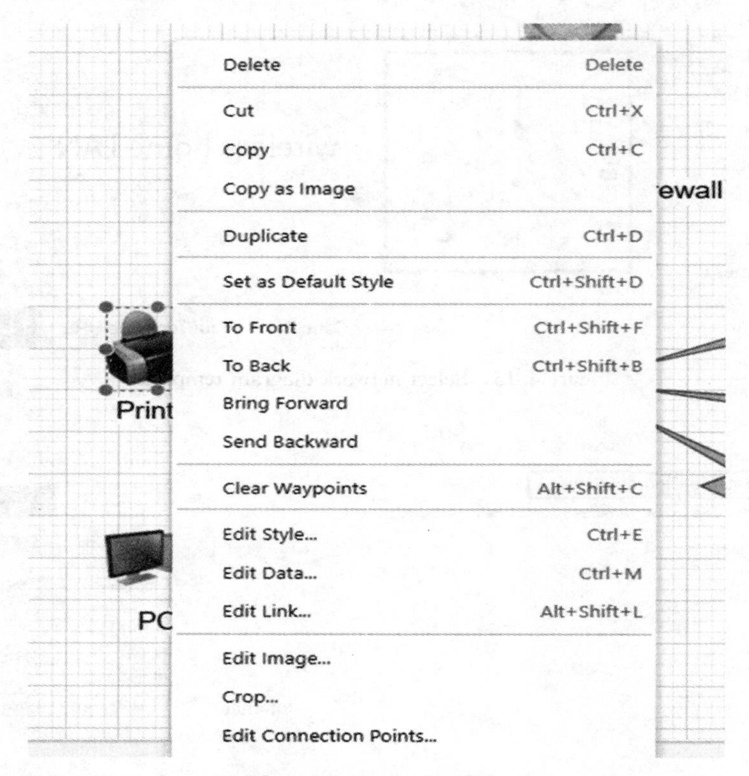

Figure 4.18 Delete an element

Tip: You can also hit the Delete button on your keyboard.

4. Add network components to the diagram

In the left column of the screen, you'll notice a plethora of computer-related icons. You can choose from our options, standard icons, Cisco (basic and extended), network, electronics, audio equipment, and other images. There are a lot of options, so you may want to enter the name of the network device you're looking for in the search box at the top of the left column. You can also scroll through the images/icons. When you see one you like, click on it and drag it to the screen (See Figure 4.19).

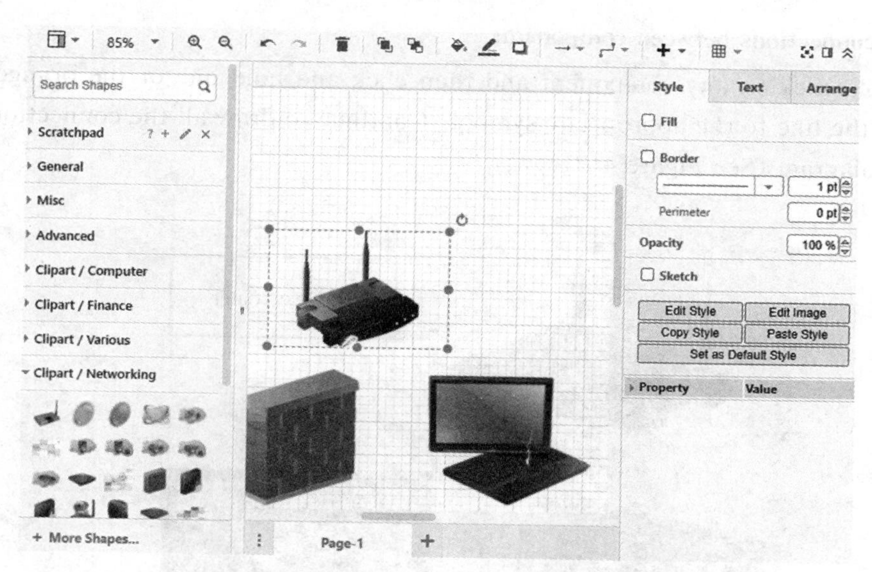

Figure 4. 19 Add network components to the diagram

5. Name the items in your network diagram

Before you start drawing network connections, let's name the items added to the diagram (See Figure 4. 20). As you can see, you can group entities by drawing squares around them. Here's how to add text and draw squares.

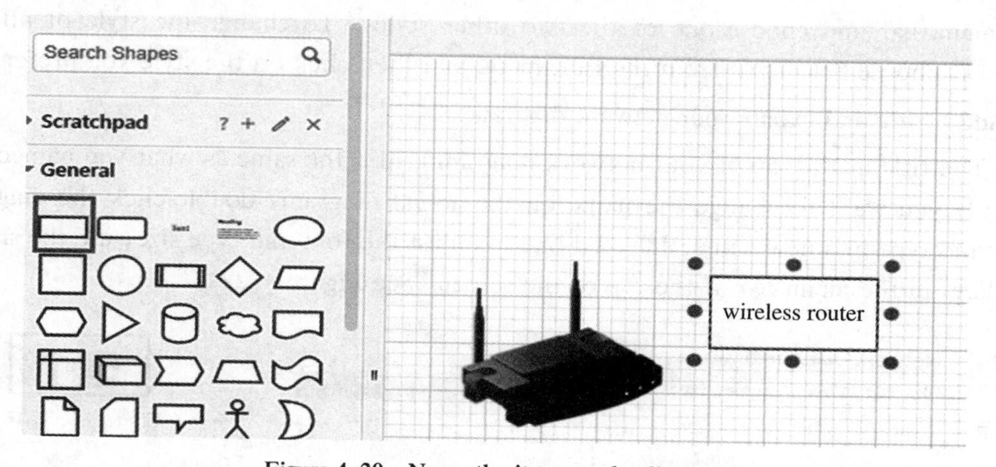

Figure 4. 20 Name the items to the diagram

Add text: To add text to the diagram, scroll through the left column and look for the heading General (you'll see a T, a square, arrow, note, and a colored box). Click on the Text and drag it to the place on the diagram where you want to add text. Type in the text and use the menu options for color, font, size, bold, etc. to customize it.

6. Draw connections between components

Double-click on any component and then click and hold one of the orange circles, and drag the line to the appropriate symbol. Continue to draw all the connections on the network diagram (See Figure 4.21).

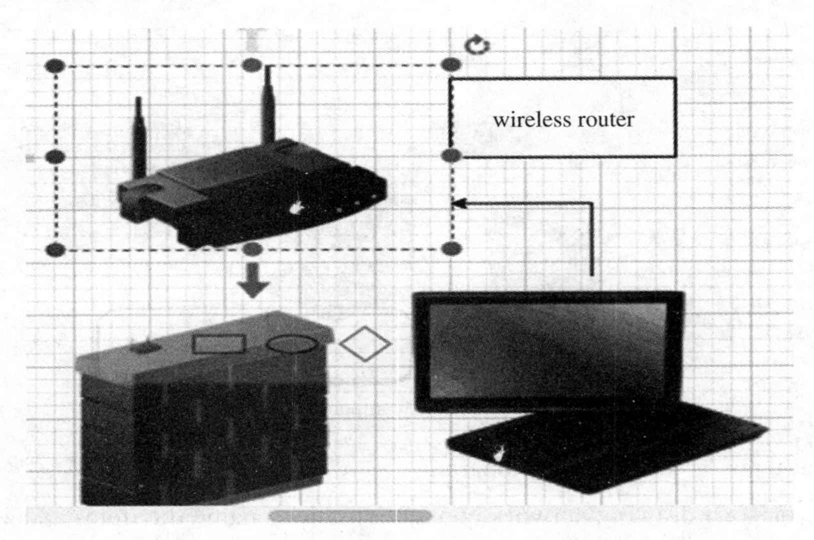

Figure 4.21 Draw connections between components

Tip: You can adjust the arrow style by clicking on it. Next, click the arrow icon in the menu bar and choose one of the two other styles. To change the style of all the arrows, choose Select All from the Edit menu and then click on the style you prefer.

7. Add a title and export your network diagram

The title at the top of your network diagram grid is the same as what you named the file. If you want to change the name on the actual diagram, double-click the diagram title and type in a new name. If you'd like to adjust the font and type size, use the shortcut keys in the menu bar at the top of the screen (See Figure 4.22).

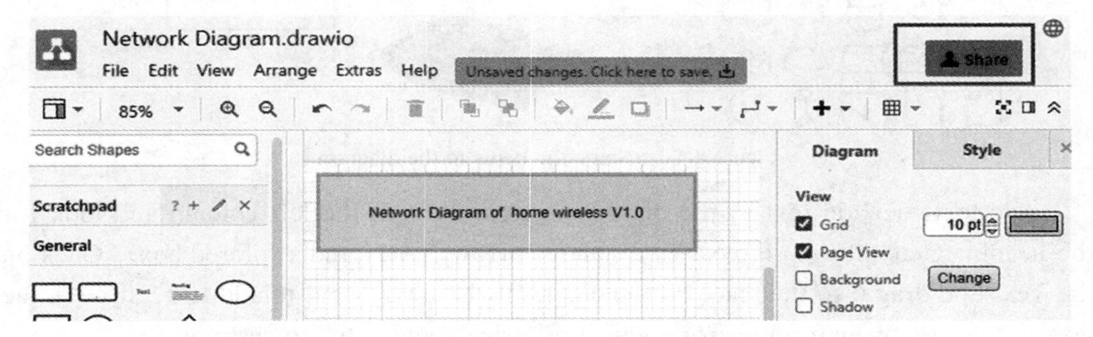

Figure 4.22 Stylean item

You can easily export your network diagram to multi-format. Click the file menu in the top left corner of the screen and a pop-up will appear (See Figure 4. 23). Choose how you'd like to export your diagram and select the appropriate information.

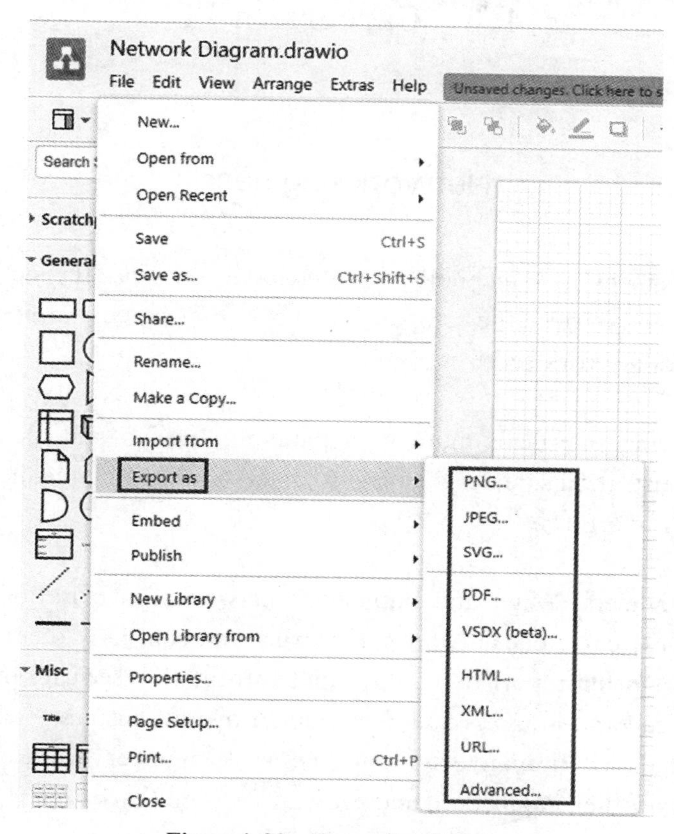

Figure 4. 23 Export the diagram

8. The simplest way to draw a network diagram

There you have it. With diagrams. net's (diagrams) intuitive and powerful online tool, it's easy to turn paper drawings into beautiful network diagrams that you'll want to share with all your coworkers. Since diagrams is a web-based App, your diagram is automatically saved, and you can access it virtually anywhere you have an Internet connection. When you share your network diagram with colleagues, you'll never have to wonder if they can open the file or access it. Give them permission to edit the diagram and they can make changes and add comments. Sign up for a free trial and see how easy it is to use diagrams.

4.3 Careers in IT

Network engineer

Network engineer is a technology professional who has the necessary skills to plan, implement and oversee the computer networks that support in-house voice, data, video and wireless network services.

The engineering side deals more with planning, design and technical specifications. The administration side deals mostly with day-to-day maintenance, management and troubleshooting efforts.

Network engineers design and implement network configurations, troubleshoot performance issues, carry out network monitoring and configure security systems such as firewalls. They often report to a CIO, chief information security officer and other line-of-business leaders to discuss and decide upon overall business goals, policies and network status updates. In many situations, network engineers work closely with project managers and other engineers, manage capacity and carry out remote or on-site support.

Some of the more popular network engineer certifications include:

- Cisco Certified Technician (CCT) Routing & Switching
- Cisco Certified Network Associate (CCNA)
- Cisco Certified Network Professional (CCNP)
- Cisco Certified Internet work Expert (CCIE)
- VMware Certified Professional — Network Virtualization (VCP-NV)

Network administrators

Network administrators manage a company's LAN and WAN networks.

They may be responsible for design, implementation, and maintenance of networks. Duties usually include maintenance of both hardware and software related to a company's intranet and Internet networks. Network administrators are typically responsible for diagnosing and repairing problems with these networks. Some network administrators' duties include planning and implementation of network security as well.

Employers typically look for candidates with a bachelor's or an associate's degree in computer science, computer technology, or information systems as well as practical networking experience. Experience with network security and maintenance is preferred. Also, technical certification may be helpful in obtaining this position. Because network administrators are involved directly with people in many departments, good communication skills are essential.

4.4 Words and phrases

asynchronous
backbone
bidirectional
bridge
broadband
Cisco Certified Internet work Expert (CCIE)
Cisco Certified Network Associate (CCNA)
Cisco Certified Network Professional (CCNP)
Cisco Certified Technician (CCT)
Client-Server Network
communication system
computer network
diagram
digital life
Dynamic Host Configuration Protocol (DHCP)
ethernet
gateway
hub
hypertext document
integrate with
Internet Control Message Protocol (ICMP)
internet protocol suite
Internet Protocol (IP)

internet service provider
IoT
Local Area Network (LAN)
Metropolitan Area Network (MAN)
modem
network administrator
network architecture
network engineer
network-based application
node
Open Systems Interconnection model (OSI model)
OSI transport services
Peer-to-Peer Network
Personal Area Network (PAN)
pop-up
protocol
repeater
response time
router
scalability
security
shortcut key
switch
synchronization
telepresence
topology
Transmission Control Protocol (TCP)
video conference
VMware Certified Professional-Network Virtualization (VCP-NV)
Wide Area Network (WAN)
wireless router
World Wide Web (WWW)

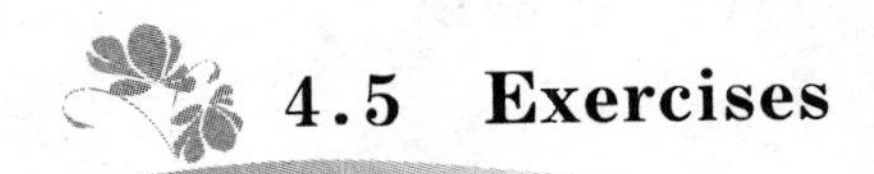

4.5 Exercises

Ⅰ. Matching

Match each numbered item with the most closely related lettered item. Write your answers in the spaces provided.

a. LAN ________(1) determining which path has been used through the subnet is what the layer should do

b. protocol ________(2) a network within a university

c. WWW ________(3) an information system where documents and other web resources are identified by uniform resource locators, which may be interlinked by hyperlinks, and are accessible over the Internet

d. TCP protocol ________(4) refers to technologies that allow a user appear to be present, feel like they are present or have some effect in a space the person does not physically inhabit

e. network layer ________(5) located in the network layer of TCP/IP model, it can provide information of various protocols to the transport layer, such as TCP, UDP, etc.

f. wan ________(6) It's a connection-oriented, reliable, byte stream-based transport layer communication protocol

g. IP address ________(7) a set of rules or procedures for transmitting data between computers

h. telepresence ________(8) a data network designed for a city

i. IP protocol ________(9) It is a numerical label such as 192.0.2.1 that is connected to a computer network that uses the Internet Protocol for communication

j. wireless router ________(10) It can be thought of as a repeater that forwards broadband Internet signals through an antenna to nearby Wi-fi devices (laptops, Wi-fi-enabled phones, tablets, and anything Wi-fi-enabled)

Ⅱ. Written practice

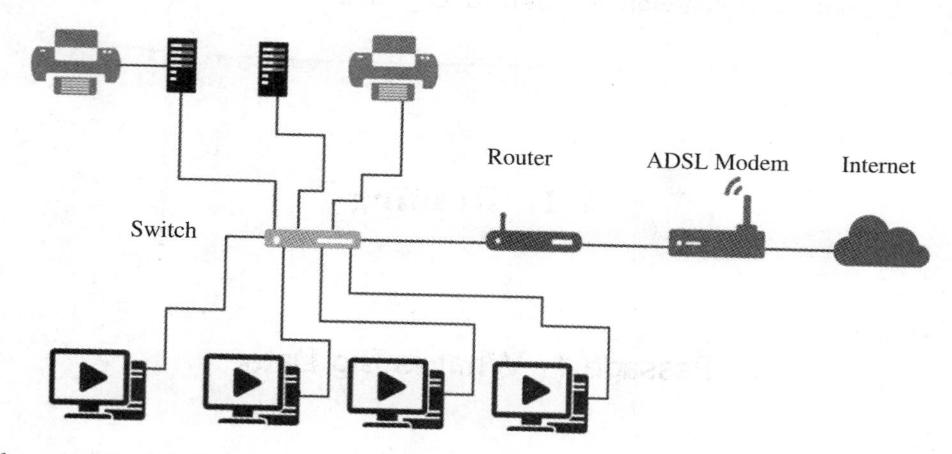

Please using the diagrams. net online tools (https://draw.io) to draw the diagram as above figure shown.

Ⅲ. Open-ended

On a separate sheet of paper, respond to each question or statement.

(1) What is 5G?

(2) Introduce the development history, architecture, address types, protocols, and advantages of ipv6.

(3) How do we assign IP addresses?

(4) Discuss network types including local area, wireless, personal, metropolitan, and wide area networks.

Chapter 5 Big Data

Learning objectives

After you have read this chapter, you should be able to:

☆ Describe the Big Data.
☆ Explain the definition and characteristics of Big Data.
☆ Understand the life cycle of Big Data process.
☆ Describe the common technology of Big Data.
☆ Understand common use cases for Big Data.

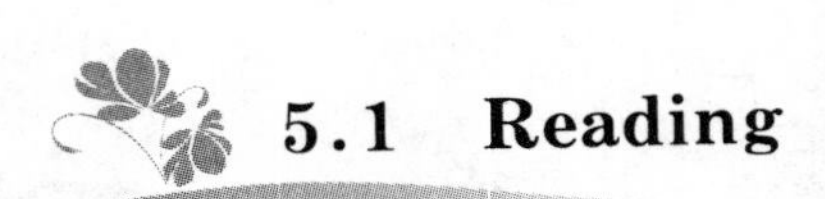

5.1 Reading

Passage 1: What is Big Data?

Ⅰ. What is Big Data?

"By 2020, it's estimated that 1.7 MB of data will be created every second for every person on earth. There are 2.5 quintillion bytes of data created each day at our current pace, but that pace is only accelerating with the growth of the IoT. Over the last two years alone, 90 percent of the data in the world was generated. (See Figure 5.1)"

Big Data is used to describe the massive volume of both structured and unstructured data that is so large, it is difficult to process using traditional techniques. So Big Data is just what it sounds like — a whole lot of data. The data comes from myriad sources: smartphones and social media posts; sensors, such as traffic signals and utility meters, point-of-sale terminals, consumer wearables such as fit meters, electronic health records.

Today, Big Data has become capital. Think of some of the world's biggest tech companies. A large part of the value they offer comes from their data, which they're constantly analyzing to produce more efficiency and develop new products.

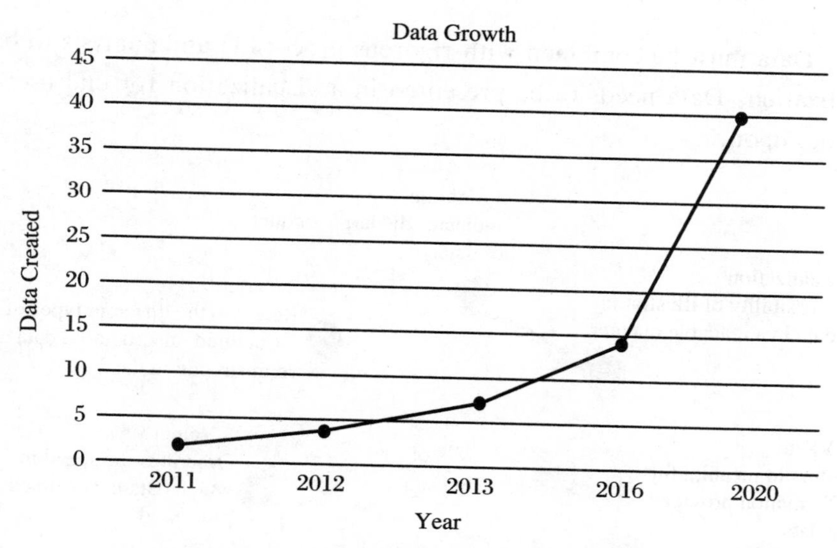

Figure 5.1 Nowadays data is growing rapidly

Recent technological breakthroughs have exponentially reduced the cost of data storage and computation, making it easier and less expensive to store more data than ever before. With an increased volume of Big Data now cheaper and more accessible, you can make more **accurate** and **precise** business decisions.

Ⅱ. What are the characteristics of Big Data?

It is important to keep in mind that Big Data isn't just about the amount of data we're generating, it's also about all the different types of data (text, video, search logs, sensor logs, customer transactions, etc.). When thinking about Big Data, consider the "seven V's" (See Figure 5.2):

Volume: Big Data is, well ... big! With the dramatic growth of the internet, mobile devices, social media, and IoT technology, the amount of data generated by all these sources has grown accordingly.

Variety: In earlier times, most data types could be neatly captured in rows on a structured table. In the Big Data world, data often comes in unstructured formats like social media posts, server log data, Lat-long Geo-coordinates, photos, audio, video and free text.

Velocity: In addition to getting bigger, the generation of data and organizations' ability to process it is accelerating.

Veracity: With many different data types and data sources, data quality issues invariably pop up in Big Data sets. Veracity deals with exploring a data set for data quality and systematically cleansing that data to be useful for analysis.

Variability: The meaning of words in unstructured data can change based on context.

Value: Data must be combined with rigorous processing and analysis to be useful.

Visualization: Data needs to be presented in a visualization for end users to understand and act upon.

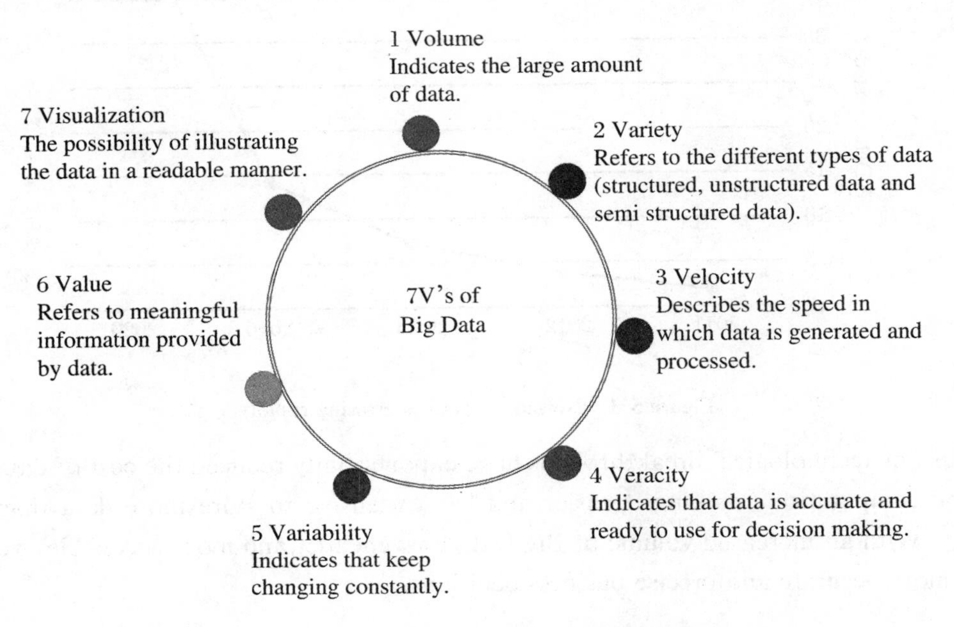

Figure 5.2 7 characteristics of Big Data

Passage 2: Exploring Big Data

Ⅰ. Big Data Process Life Cycle

On the Internet, a multitude of data is generated daily and massively. To give you a rough idea, approximately 40,000 searches are done per second on Baidu, everything we do on the internet is connected to the WWW that generates data and we call it Big Data. But the problem here is how can companies analyze these data and convert it to business-relevant information?

No two data projects are identical; each brings its own challenges, opportunities, and potential solutions that impact its trajectory. Nearly all data projects, however, follow the same basic life cycle from start to finish. The life cycle can be split into eight common stages, steps, or phases (See Figure 5.3).

Below is a walk through of the process that are typically involved in each of them.

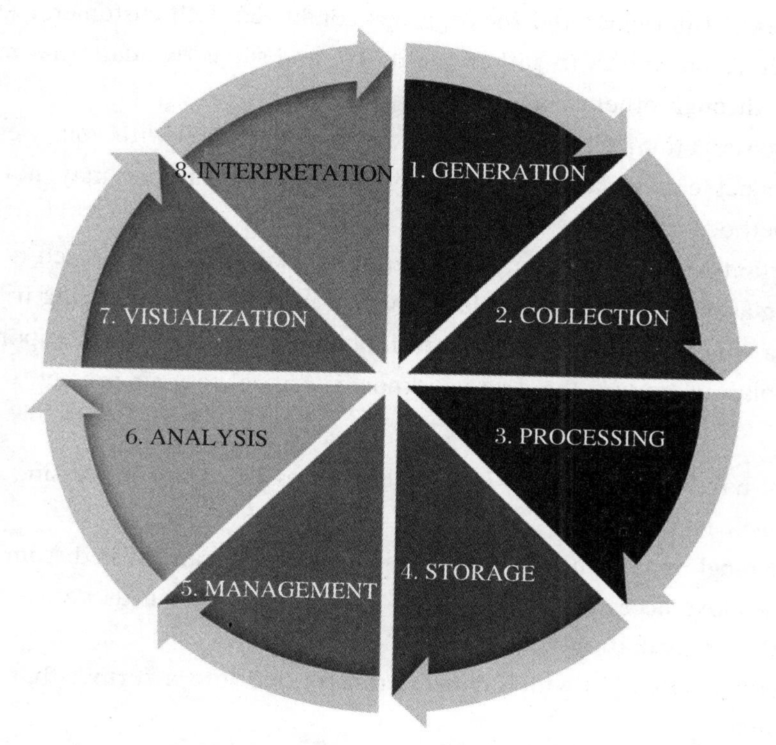

Figure 5. 3 Life cycle of Big Data process

1. Generation

For the data life cycle to begin, data must first be generated. Otherwise, the following steps can't be initiated.

Data generation occurs regardless of whether you're aware of it, especially in our increasingly online world. Some of this data is generated by your organization, some by your customers, and some by third parties you may or may not be aware of. Every sale, purchase, hire, communication, interaction — everything generates data. Given the proper attention, this data can often lead to powerful insights that allow you to better serve your customers and become more effective in your role.

2. Collection

Not all of the data that's generated every day is collected or used. It's up to your data team to identify what information should be captured and the best means for doing so, and what data is unnecessary or irrelevant to the project at hand.

You can collect data in a variety of ways, including:

• **Forms**: Web forms, client or customer intake forms, vendor forms, and human resources applications are some of the most common ways businesses generate data.

• **Surveys**: Surveys can be an effective way to gather vast amounts of information from a large number of respondents.

• **Interviews**: Interviews and focus groups conducted with customers, users, or job applicants offer opportunities to gather qualitative and subjective data that may be difficult to capture through other means.

• **Direct observation**: Observing how a customer interacts with your website, application, or product can be an effective way to gather data that may not be offered through the methods above.

It's important to note that many organizations take a broad approach to data collection, capturing as much data as possible from each interaction and storing it for potential use. While drawing from this supply is certainly an option, it's always important to start by creating a plan to capture the data you know is critical to your project.

3. Processing

Once data has been collected, it must be processed. Data processing can refer to various activities, including:

• **Data wrangling**, in which a data set is cleaned and transformed from its raw form into something more accessible and usable. This is also known as data cleaning, data munging, or data remediation.

• **Data compression**, in which data is transformed into a format that can be more efficiently stored.

• **Data encryption**, in which data is translated into another form of code to protect it from privacy concerns.

Even the simple act of taking a printed form and digitizing it can be considered a form of data processing.

4. Storage

After data has been collected and processed, it must be stored for future use. This is most commonly achieved through the creation of databases or datasets. These datasets may then be stored in the cloud, on servers, or using another form of physical storage like a hard drive, CD, cassette, or floppy disk.

When determining how to best store data for your organization, it's important to build in a certain level of redundancy to ensure that a copy of your data will be protected and accessible, even if the original source becomes corrupted or compromised.

5. Management

Data management, also called database management, involves organizing, storing, and retrieving data as necessary over the life of a data project. While referred to here as a "step", it's an ongoing process that takes place from the beginning through the end of a project. Data management includes everything from storage and encryption to implementing access logs and change logs that track who has accessed data and what changes they may have made.

6. Analysis

Data analysis refers to processes that attempt to glean meaningful insights from raw data. Analysts and data scientists use different tools and strategies to conduct these analyses. Some of the more commonly used methods include statistical modeling, algorithms, artificial intelligence, data mining, and machine learning.

Exactly who performs an analysis depends on the specific challenge being addressed, as well as the size of your organization's data team. Business analysts, data analysts, and data scientists can all play a role.

7. Visualization

Data visualization refers to the process of creating graphical representations of your information, typically through the use of one or more visualization tools (See Figure 5.4). Visualizing data makes it easier to quickly communicate your analysis to a wider audience both inside and outside your organization. The form your visualization takes depends on the data you're working with, as well as the story you want to communicate.

While technically not a required step for all data projects, data visualization has become an increasingly important part of the data life cycle.

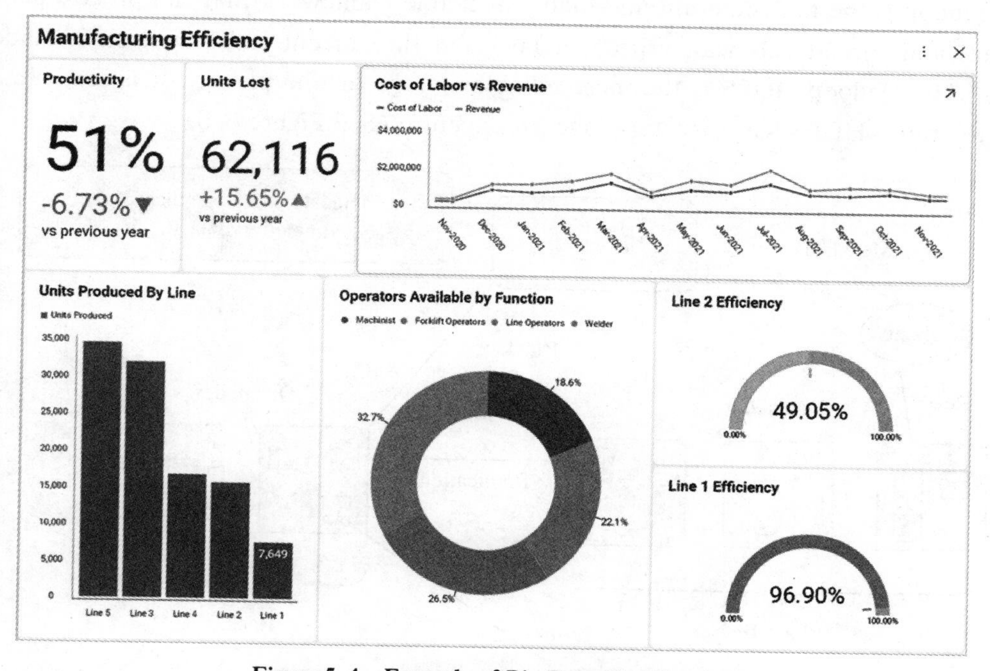

Figure 5.4 Example of Big Data visualization

8. Interpretation

Finally, the interpretation phase of the data life cycle provides the opportunity to make sense of your analysis and visualization. Beyond simply presenting the data, this is when you investigate it through the lens of your expertise and understanding. Your interpretation may not only include a description or explanation of what the data shows, but more importantly, what the implications may be.

Ⅱ. Big Data technologies

Big Data technologies are the software utility designed for analyzing, processing, and extracting information from the unstructured large data which can't be handled with the traditional data processing software.

Companies required Big Data processing technologies to analyze the massive amount of real-time data. They use Big Data technologies to come up with predictions to reduce the risk of failure.

1. Apache Hadoop

It is the topmost big data tool. Apache Hadoop is an open-source software framework developed by Apache Software Foundation for storing and processing Big Data. Hadoop stores and processes data in a distributed computing environment across the cluster of commodity hardware.

Hadoop is the fault-tolerant and highly available framework that can process data of any size and formats. It was written in Java and the current stable version is Hadoop 3.1.3. The Hadoop HDFS is the most reliable storage on the planet. It is scalable and fault-tolerant. HDFS has a master/slave architecture (See Figure 5.5).

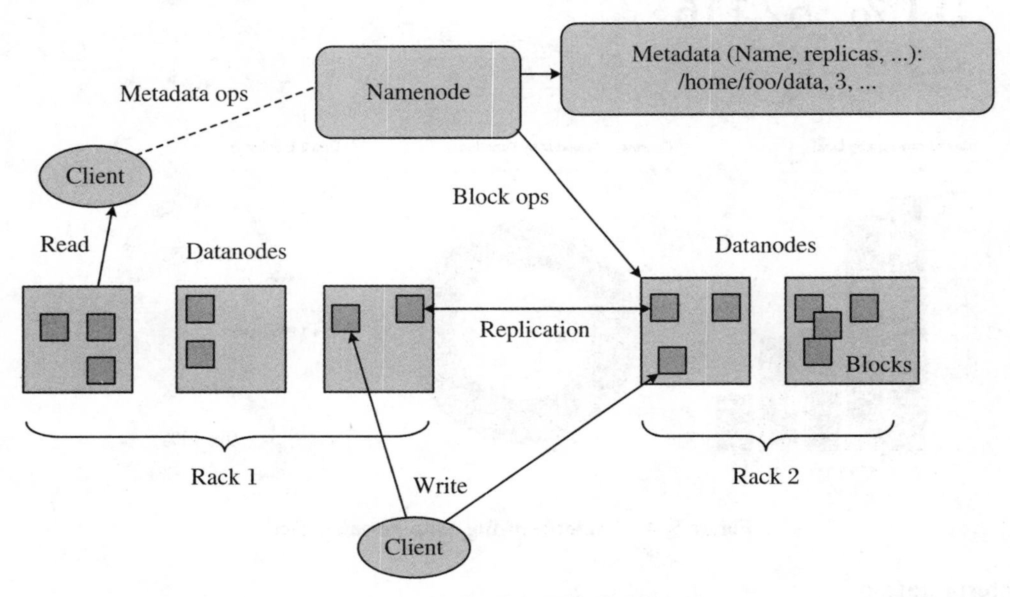

Figure 5.5　HDFS architecture

The framework is designed in such a way that it can work even in unfavorable conditions like a machine crash.

The framework stores data across commodity hardware that makes Hadoop cost-effective.

Hadoop stores and processes data in a distributed manner. The data is processed parallelly resulting in fast data processing.

Companies using Hadoop are Facebook, LinkedIn, IBM, Intel, Microsoft, and many more.

2. Apache Spark

Apache Spark is another popular open-source Big Data tool designed with the goal to speed up the Hadoop Big Data processing. The main objective of the Apache Spark Project was to keep the advantages of MapReduce's distributed, scalable, fault-tolerant processing framework and make it more efficient and easier to use.

It provides in-memory computing capabilities to deliver speed. Spark supports both real-time as well as batch processing and provides high-level APIs in Java, Scala, Python, and R.

Features of Apache Spark:

- Spark has the ability to run applications in Hadoop clusters 100 times faster in memory and ten times faster on the disk.
- Apache Spark can work with different data stores (such as OpenStack, HDFS, Cassandra) due to which it provides more flexibility than Hadoop.
- Spark contains a ML library that offers a dynamic group of machine algorithms such as Clustering, Collaborative, Filtering, Regression, Classification, etc.

Batch/streaming data

Unify the processing of your data in batches and real-time streaming, using your preferred language: Python, SQL, Scala, Java or R.

SQL analytics

Execute fast, distributed ANSI SQL queries for dashboarding and ad-hoc reporting. Runs faster than most data warehouses.

Data science at scale

Perform Exploratory Data Analysis (EDA) on petabyte-scale data without having to resort to downsampling.

Machine learning

Train machine learning algorithms on a laptop and use the same code to scale to fault-tolerant clusters of thousands of machines.

Figure 5.6 Key features of Apache Spark

• Apache Spark can run on Hadoop, Kubernetes, Apache Mesos, standalone, or in the cloud.

3. MongoDB

MongoDB is an open-source document-oriented database written in C, C++, and JavaScript and has an easy setup environment.

MongoDB is one of the most popular databases for Big Data. It facilitates the management of unstructured or semi-structured data or the data that changes frequently.

MongoDB executes on MEAN software stack, NET applications, and Java platforms. It is also flexible in cloud infrastructure.

Features of MongoDB:

• It is highly reliable, as well as cost-effective.

• It has a powerful query language that provides support for aggregation, geo-based search, text search, graph search.

• Supports ad hoc queries, indexing, sharing, replication, etc.

• It has all the powers of the relational database.

Companies like Facebook, eBay, MetLife, Google, etc. use MongoDB.

4. Apache Kafka

Apache Kafka is an open-source distributed streaming platform developed by Apache Software Foundation. It is a publish-subscribe based fault-tolerant messaging system and a robust queue capable of handling large volumes of data. It allows us to pass the message from one point to another. Kafka is used for building real-time streaming data pipelines and real-time streaming applications. Kafka is written in Java and Scala. Apache Kafka integrates very well with Spark and Storm for real-time streaming data analysis.

Features of Apache Kafka:

• Kafka can work with huge volumes of data easily.

• Kafka is highly scalable, distributed and fault-tolerant.

• It has high through put for both publishing and subscribing messages.

• It guarantees zero downtime and no data loss.

5. Apache Storm

It is a distributed real-time computational framework. Apache Storm is written in Clojure and Java. With Apache Storm, we can reliably process our unbounded streams of data. It is a simple tool and can be used with any programming language.

We can use Apache Storm in real-time analytics, continuous computation, online machine learning, ETL, and more.

Features of Apache Storm:

• It is free and open-source.

• It is highly scalable.

- Apache Storm is fault-tolerant and easy to set up.
- Apache Storm guarantees data processing.
- It has the capability to process millions of tuples per second per node.

Companies like Yahoo, Alibaba, Groupon, Twitter, Spotify use Apache Storm.

6. Apache Hive

Hive is an open-source data warehousing tool for analyzing Big Data. Hive uses Hive Query Language (HQL) which is similar to SQL for querying unstructured data.

It is built on the top of Hadoop and enables developers to perform processing on data stored in Hadoop HDFS without writing the complex MapReduce jobs. Users can interact with Hive through CLI (Beeline Shell).

Features of Apache Hive:

- Hive provides support for all the client applications.
- It reduces the overhead of writing complex MapReduce jobs.
- HQL syntax is similar to SQL. Thus, one who is familiar with SQL can easily write Hive queries.

7. Tableau

Tableau is a powerful data visualization and software solution tools in the Business Intelligence (BI) and analytics industry.

It is the perfect tool for transforming the raw data into an easily understandable format without any technical skill and coding knowledge.

Tableau allows you to work on the live datasets and turns the raw data into valuable insights and enhances the decision-making process.

It offers a rapid data analysis process, which results in visualizations that are in the form of interactive dashboards and worksheets. It works in synchronization with the other Big Data tools.

Features of Tableau:

- In Tableau, with simple drag and drop, one can make visualizations in the form of a Bar chart, Pie chart, Histogram, Boxplot, Gantt chart, Bullet chart, and many more.
- Tableau offers a large option of data sources ranging from on-premise files, Text files, CSV, Excel, relational databases, spreadsheets, non-relational databases, Big Data, data warehouses, to on-cloud data.
- It is highly robust and secure.
- It allows the sharing of data in the form of visualizations, dashboards, sheets, etc. in real-time.

Ⅲ. Big Data analytics use cases

1. Big Data in education industry

The education industry is flooded with huge amounts of data related to students, faculty, courses, results, and whatnot (See Figure 5.7). Now, we have realized that proper study and analysis of this data can provide insights that can be used to improve the operational effectiveness and working of educational institutes.

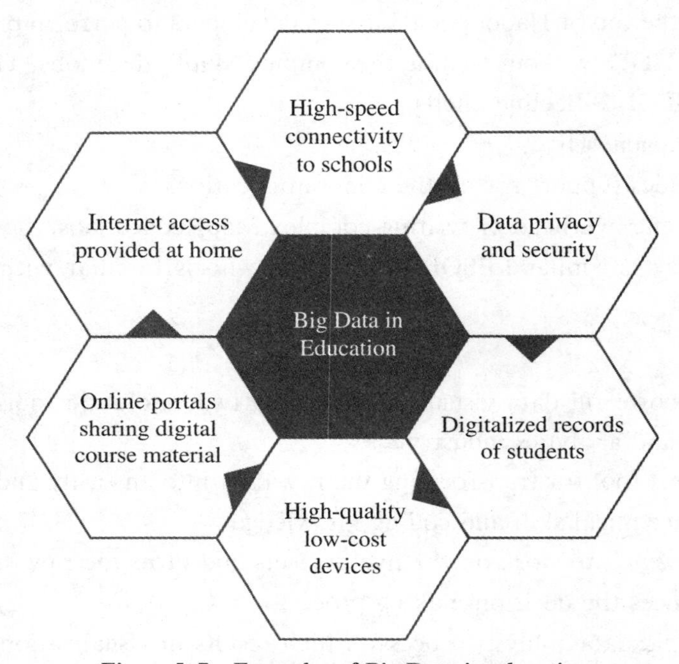

Figure 5.7 Examples of Big Data in education

Following are some of the fields in the education industry that has been transformed by Big Data motivated changes:

Customized and dynamic learning programs

Customized programs and schemes to benefit individual students can be created using the data collected based on each student's learning history. This improves the overall student results.

Reframing course material

Reframing the course material according to the data that is collected based on what a student learns and to what extent by real-time monitoring of the components of a course is beneficial for the students.

Grading systems

New advancements in grading systems have been introduced as a result of a proper analysis of student data.

Career prediction

Appropriate analysis and study of every student's records will help understand each student's progress, strengths, weaknesses, interests, and more. It would also help in determining which career would be the most suitable for the student in the future.

The applications of Big Data have provided a solution to one of the biggest pitfalls in the education system, that is, the one-size-fits-all fashion of academic set-up, by contributing to e-learning solutions.

2. Big Data in healthcare industry

Healthcare is yet another industry that is bound to generate a huge amount of data. Following are some of how Big Data has contributed to healthcare (See Figure 5. 8):

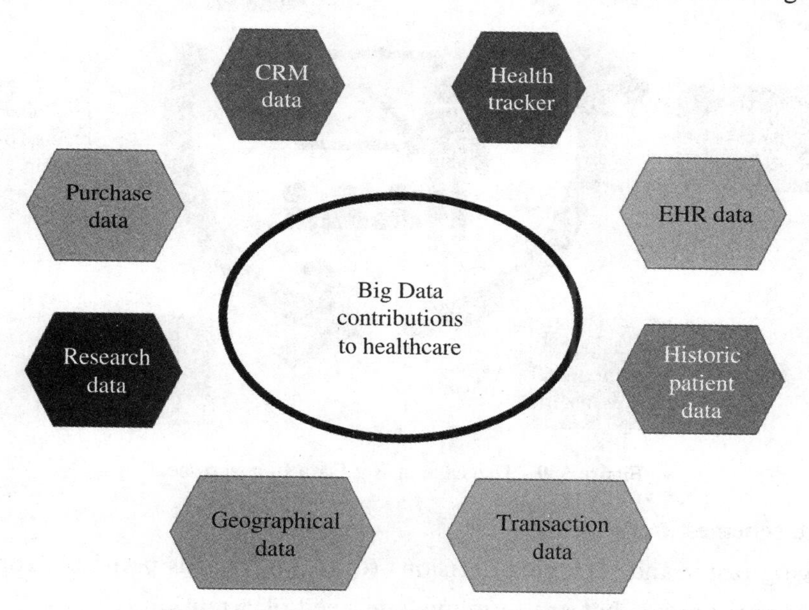

Figure 5. 8 Use case of Big Data in healthcare

- Big Data reduces the costs of a treatment since there are fewer chances of having to perform unnecessary diagnoses.
- It helps in predicting outbreaks of epidemics and also in deciding what preventive measures could be taken to minimize the effects of the same.
- It helps avoid preventable diseases by detecting them in the early stages. It prevents them from getting any worse which in turn makes their treatment easy and effective.
- Patients can be provided with evidence-based medicine which is identified and prescribed after researching past medical results.

3. Big Data in government sector

Government comes face to face with a very huge amount of data on an almost daily basis. They have to keep track of various records and databases regarding their citizens, their growth, energy resources, geographical surveys, and many more (See Figure 5.9). All the data contributes to Big Data. The proper study and analysis of this data, hence, helps governments in endless ways. A few of them are as follows:

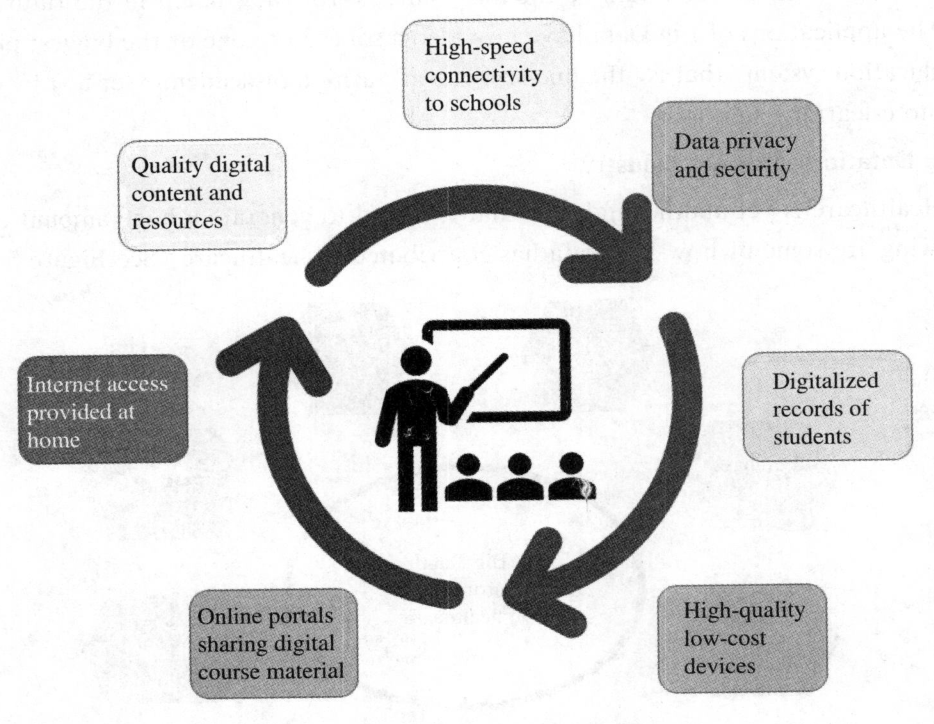

Figure 5.9 Use case of Big Data in government

Welfare schemes

In making faster and informed decisions regarding various political programs:

- To identify areas that are in immediate need of attention.
- To stay up to date in the field of agriculture by keeping track of all existing land and livestock.
- To overcome national challenges such as unemployment, terrorism, energy resources exploration, and much more.

Cyber security

- Big Data is hugely used for deceit recognition in the domain of cyber security.
- It is also used in catching tax evaders.
- Cyber security engineers protect networks and data from unauthorized access.

5.2 Writing: How to write a thank-you letter

Thank-you letter

There's an art to writing a thank-you letter. It goes beyond saying, "Thanks for ________. I really appreciate it." We'll show you some thank-you letter examples and templates that will help you express your gratitude in style.

What to include in a thank-you letter?

No matter what form you use to send your thank-you note, there are certain components you should always include.

1. Address the person appropriately

At the start of the letter, address the person with a proper salutation, such as "Dear Mr. last name." or "Dear first name." If you know the person well, use the person's first name. Otherwise, address him or her as Mr., Ms., or another appropriate title.

2. Say thank you

Get to the point of your note quickly. Say the words "thank you" in the first sentence or two, so the person knows why you are writing. If you are sending an email, include the phrase "thank you" in the subject line as well.

3. Give (some) specifics

Make sure you specify what you are thanking the individual for. Go into a bit of detail, so the person understands exactly what you appreciate. For example, if you are saying thank you to someone who gave you job advice, explain what you found to be most helpful. If you are saying thank you after a job interview, remind the person of a particular moment from the interview (or remind them why you are a good fit for the job).

4. Say thank you again

Before signing off, reiterate your appreciation.

5. Sign off

Use an appropriate closing, such as "Best" or "Sincerely." Then end with your signature (handwritten and typed if it is a letter, handwritten if it is a card, and typed if it is an email).

Sample

Dear Ms. Kingston,

Thank you for taking the time to meet with me yesterday to chat about the content marketing manager position at Really Big Corporation. It's my pleasure connecting with you and hearing how energized you are about the company's content marketing and growth goals. Because of my background in retail marketing, I was particularly interested in your innovative ideas for retail outreach — they sparked some ideas of my own and left me with the sense that we'd make an excellent collaborative team.

You mentioned that you'll be taking some time to make a hiring decision, so I'll do my best to wait patiently despite how excited I am to be considered. Meanwhile, let me know if there's any further info I can provide. Thanks again for choosing me.

All the best,

David

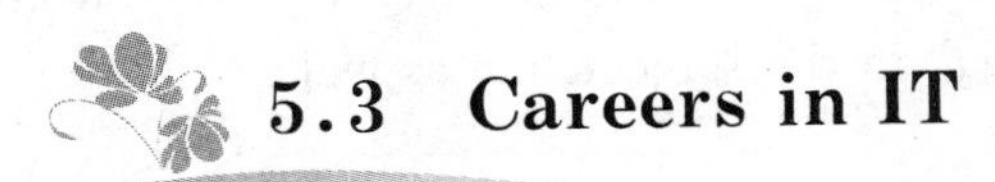

5.3 Careers in IT

Big Data developer

The roles and responsibilities of the Big Data developer who is responsible for programming Hadoop applications in the Big Data fields are described as following:

- Loading the data from disparate data sets.
- High-speed querying.
- Propose best practices and standards.
- Your role maybe to design, build, install, configure and support Hadoop.
- Maintains security and data privacy.
- Manages and deploys HBase.

• Performs an analysis of the vast number of data stores and uncovers insights.

• Big Data developer is responsible for Hadoop development and implementation.

• He is responsible for creating scalable and high-performance web services for tracking data.

• Translates complex technical and functional requirements into detailed designs.

• He proposes design changes and suggestions to various processes and products.

Big Data architect

The role of a data architect is that of a visionary leader in an organization. A data architect's roles and responsibilities including:

• Developing and implementing an overall organizational data strategy that is in line with business processes. The strategy includes data model designs, database development standards, implementation and management of data warehouses and data analytics systems.

• Identifying data sources, both internal and external, and working out a plan for data management that is aligned with organizational data strategy.

• Coordinating and collaborating with cross-functional teams, stakeholders, and vendors for the smooth functioning of the enterprise data system.

• Managing end-to-end data architecture, from selecting the platform, designing the technical architecture, and developing the application to finally testing and implementing the proposed solution.

• Planning and execution of Big Data solutions using technologies such as Hadoop. In fact, the Big Data architect roles and responsibilities entail the complete life-cycle management of a Hadoop solution.

• Defining and managing the flow of data and dissemination of information within the organization.

• Integrating technical functionality, ensuring data accessibility, accuracy, and security.

• Conducting a continuous audit of data management system performance, refine whenever required, and report immediately any breach or loopholes to the stakeholders.

Big Data engineer

The main responsibilities of a Big Data engineer are the following:

- Gathering and processing raw data and translating analyses.
- Evaluating new data sources for acquisition and integration.
- Designing and implementing relational databases for storage and processing.
- Working directly with the technology and engineering teams to integrate data processing and business objectives.

5.4 Words and phrases

accurate
algorithm
Apache Software Foundation
architect
characteristics
cluster
collaborate
commodity hardware
coordinate
create
data compression
data encryption
data mining
data wrangling
description and explanation
distributed computing
document-oriented database
environment
exponentially
facilitate
fault-tolerant
features
form
framework
generated
generation
glean
HDFS
highly available
Hive Query Language (HQL)
huge
infrastructure
interpretation
interview
massive
MongoDB
myriad
open-source
parallelly
precise
real-time
reduce the risk of failure
redundancy
retrieve

semi-structured
speed up
stable version
survey
Tableau
variability
variety
velocity
veracity
visualization
volume

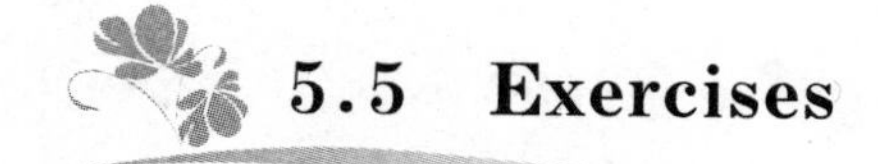

5.5 Exercises

Ⅰ. Matching

Match each numbered item with the most closely related lettered item. Write your answers in the spaces provided.

a. volume
b. velocity
c. variety
d. veracity
e. variability
f. visualization
g. collection
h. management
i. analysis

________(1) the fast rate at which data is received and processed

________(2) the types of data that are available, like text, audio, video

________(3) creating graphical representations of your information, typically through the use of one or more tools

________(4) the amount of data matters, contains low-density, unstructured data

________(5) explore a data set for data quality and systematically cleansing that data to be useful for analysis

________(6) unstructured data can change based on context

________(7) get the data to identify what information should be captured and the best means for doing so, by Forms, Surveys, Interviews and so on

________(8) attempt to glean meaningful insights from raw data

________(9) organize, store, and retrieve data as necessary over the life of a data project

Ⅱ. Written practice

Thanking friends and family

Sometimes, we forget to thank the people closest to us for the things they do or give to us. When a heartfelt face-to-face thank-you isn't possible, a brief letter, card, or email is an excellent way to show that your friends' and family members' contributions haven't gone unnoticed.

Ⅲ. Open-ended

On a separate sheet of paper, respond to each question or statement.

(1) What Big Data applications have you encountered in your daily life?
(2) What's your view about the other use cases of Big Data?
(3) Consider the challenges of Big Data biology.

Chapter 6 AI and machine learning

Learning objectives

After you have read this chapter, you should be able to:

- ☆ Explain AI, ML and DL.
- ☆ Explain stages of AI, including ANI, AGI and ASI.
- ☆ Describe the types of ML, including supervised learning, unsupervised learning and reinforcement learning.
- ☆ List some examples of AI in every area of our lives.
- ☆ Describe the relationship among AI, ML and DL.
- ☆ Distinguish the critical differences between traditional programming and ML.

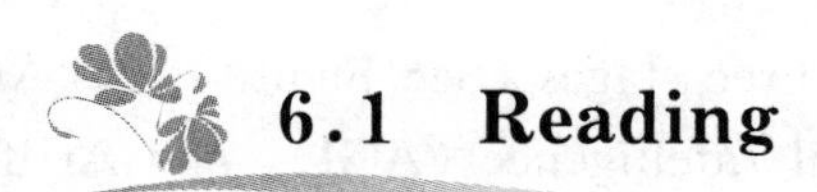

6.1 Reading

Passage 1: Introduction to artificial intelligence

In March 2016, AlphaGo beat Lee Sedol in a five-game match, the first time a computer Go program has beaten a 9-duan professional without handicap, the first to defeat a Go world champion (See Figure 6.1). Go is a very difficult game for computers to play, and has long been considered one of the grand challenges of AI.

Ⅰ. What is AI?

Artificial Intelligence (AI) is a field that has a long history, and known as machine intelligence. The term was coined in 1956 by John McCarthy, a researcher who later founded AI labs at MIT and Stanford. It is a branch of computer science that focuses on building and managing technology that can learn to autonomously make decisions and carry out actions on behalf of a human being.

Figure 6.1　AlphaGo VS Lee Sedol

AI is not a single technology. It is an umbrella term that includes any type of software or hardware component that supports machine learning. It includes the following technologies: Computer Vision (CV), Natural Language Processing (NLP), speech processing, expert systems, etc.

Ⅱ. What are the stages of AI?

AI is divided broadly into three stages (See Figure 6.2): Artificial Narrow Intelligence (ANI), Artificial General Intelligence (AGI), and Artificial Super Intelligence (ASI). ANI is considered "weak" AI, whereas the other two types are classified as "strong" AI.

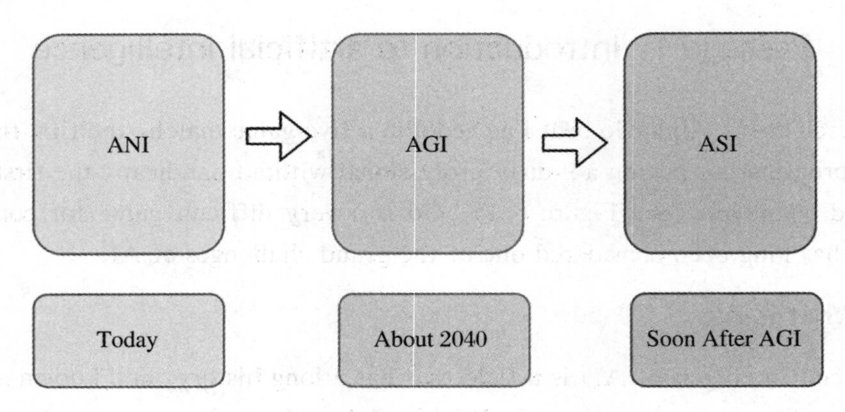

Figure 6.2　AI stages

Weak AI is defined by its ability to complete a very specific task, like winning a chess game. Strong AI is defined by its ability to compare to humans. AGI would perform on par with another human while ASI — also known as super intelligence — would surpass a human's intelligence.

Ⅲ. How is AI impacting our lives?

AI assists in every area of our lives (See Figure 6. 3). Here is a list of examples of AI you're likely using in daily day:

- Short video recommendation of Tik Tok.
- IFLYTEK speech recognition.
- Voice-to-text features of WeChat.
- Conversational bots of Taobao.
- Spam filters on email, etc.

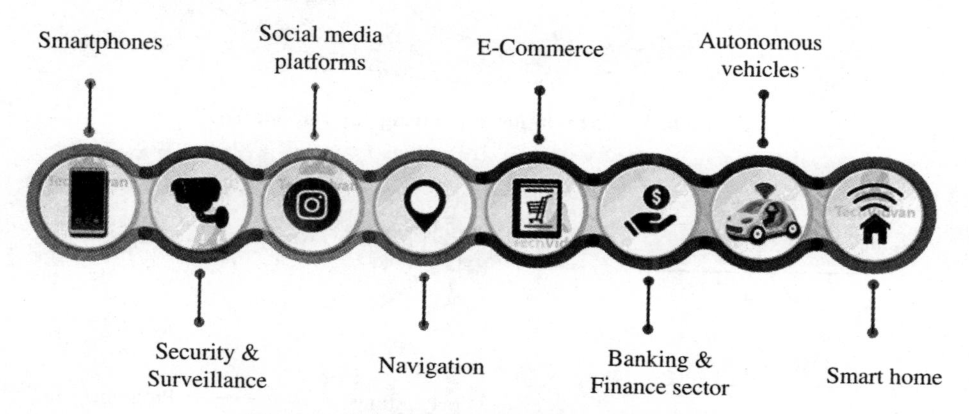

Figure 6. 3 Scenarios of artificial intelligence

Although these are some examples of weak AI, AI is still constantly and rapidly growing and changing. A lot of AI researchers believe strong AI can be achieved in the future, and predict that the widespread use of AI will lead to a second industrial revolution. Artificial intelligence has already made a big impact on many aspects of our day-to-day tasks. As more advancements are made to AI technology, it will become even more prevalent in daily life and play an even more fundamental role in our society.

Passage 2: Machine learning and deep learning

Ⅰ. What is machine learning?

ML (machine learning) is an application of AI that provides systems the ability to learn on their own and improve from experiences without being programmed externally. It is a branch of AI and computer science which focuses on the use of data and algorithms to imitate the way that humans learn, gradually improving its accuracy.

Machine learning algorithms build a model based on sample data, known as training data, in order to make predictions or decisions without being explicitly programmed to do so (See Figure 6.4 and Figure 6.5).

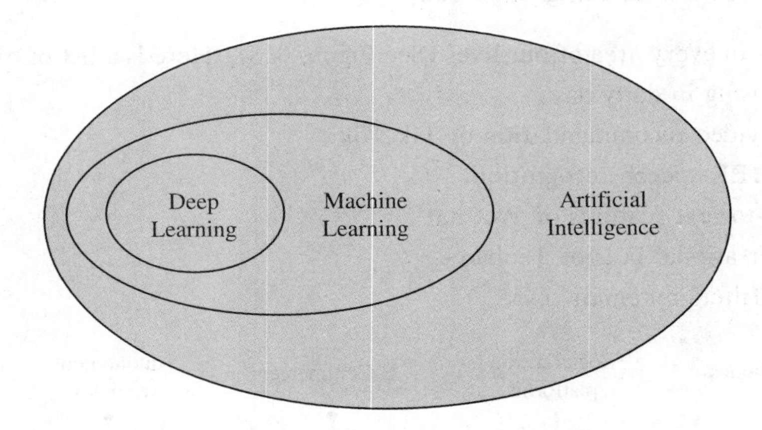

Figure 6.4 The relationship among AI, ML and DL

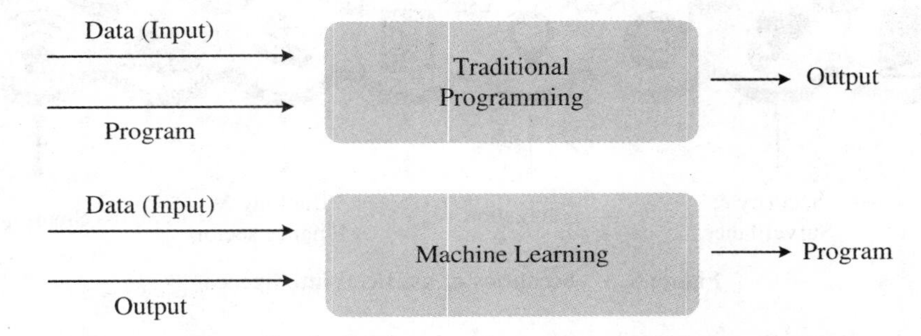

Figure 6.5 Traditional programming VS Machine learning

Ⅱ. What are the types of machine learning?

Machine learning can be classified into three types of algorithms:

- Supervised learning.
- Unsupervised learning.
- Reinforcement learning.

In supervised learning, an AI system is presented with data which is labeled, which means that each data tagged with the correct label (See Figure 6.6).

In unsupervised learning, an AI system is presented with unlabeled, uncategorized data and the system's algorithms act on the data without prior training (See Figure 6.7). The output is dependent upon the adopted algorithms.

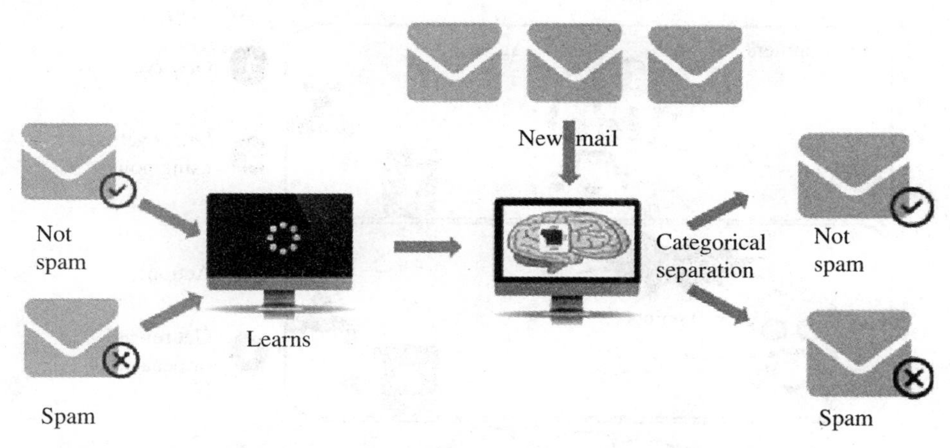

Figure 6. 6 Example of supervised learning: classify email

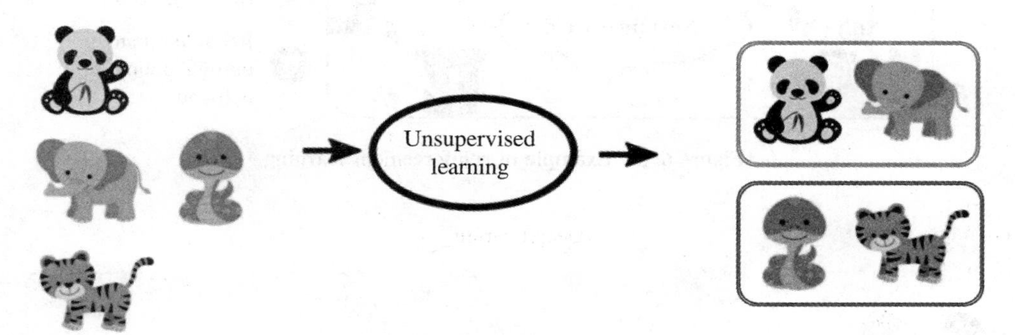

Figure 6. 7 Example of unsupervised learning

A reinforcement learning algorithm, or agent, learns by interacting with its environment. The agent receives rewards by performing correctly and penalties for performing incorrectly. It is a type of dynamic programming that trains algorithms using a system of reward and punishment (See Figure 6. 8).

Ⅲ. What is deep learning?

DL (deep learning) is a sub-field of machine learning concerned with algorithms inspired by the structure and function of the brain called Artificial Neural Networks (ANN) (See Figure 6. 9).

Deep learning concepts are used to teach machines what comes naturally to us humans. Using deep learning, a computer model can be taught to run classification acts taking image, text, or sound as an input. DL is becoming popular as the models are capable of achieving state of the art accuracy. Large labeled data sets are used to train these models along with the neural network architectures.

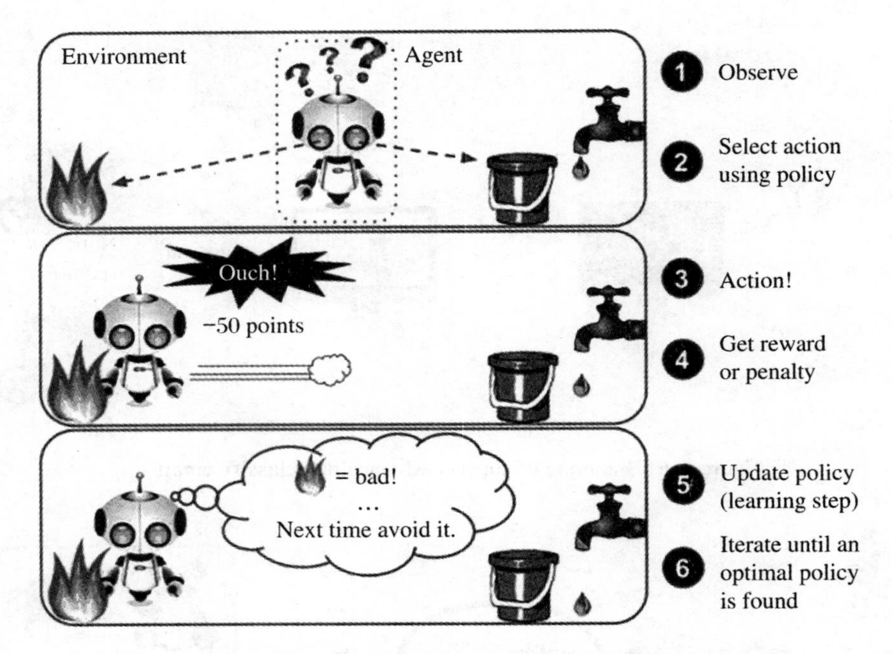

Figure 6.8 Example of reinforcement learning

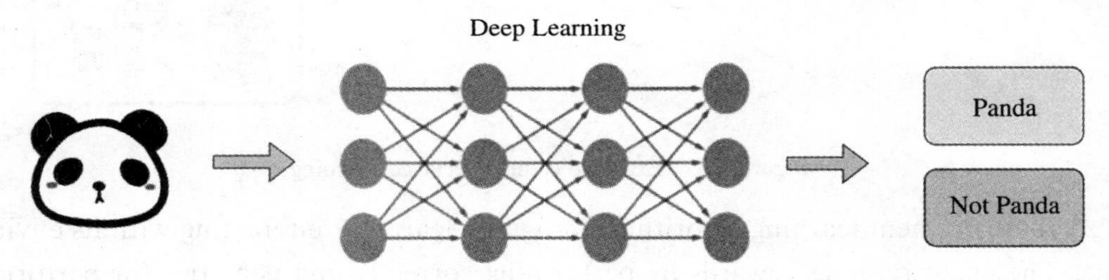

Figure 6.9 Deep learning model

Ⅳ. The future of machine learning

Machine learning is perhaps the hottest thing in the field of AI. Scientists and experts have been working to develop a computer that acts more like humans in the post-industrialized phase. Machine learning revolution will stay with us for long and so will be the future of machine learning.

6.2 Writing: How to write an invitation letter

Invitation letter

Invitation letters are those letters that are written to invite individuals to a specific event. The main purpose of writing invitation letters is to coordinate the number of guests coming a few days before the date of the event. An invitation letter helps the host handle the event better as they can make arrangements accordingly. We could be written for a conference, a wedding, a graduation ceremony, an exhibition, or an annual day, etc.

Ⅰ. Why are invitations so important

In a world that is dominated by instant messaging and Apps like WeChat. But there are some reasons why you should consider investing your time in writing letters to invite people to your party or event.

First impressions are really the last impressions. If your event is really professional, sending a text message to your guests would hardly be appropriate. You have to send out formal invitation letters to give the perception that your event is extremely professional and has to be taken seriously.

Ⅱ. Composing the letter or email

You should always start your invitation letter with phrases like:

- It is my great pleasure to invite you ...
- We're pleased to welcome you ...
- Our organization will be venerated to welcome you as a guest ...
- It would be a pleasure for us if you could come ...
- We would be glad if you could come to ...
- On behalf of our organization, we would like to welcome you ...

Writing phrases like these at the beginning of the letter demonstrates your respect and happiness towards inviting an individual to the event.

After you've written that, specify the intent of the event clearly in the first paragraph itself. Mention the most important details like the date, time, and venue in the first paragraph itself. It would be convenient for the recipient to find these important details without reading the entire letter again.

In the second paragraph, you should describe the event's purpose and why you believe it needs to be attended by the recipient.

In addition to all of that, more information can be attached related to the event. For instance, if there is a program for the event, it is better to mention it in the text itself. Additionally, if some special guests or events have been planned for the event, they can be listed in the letter as well.

You can provide additional instructions for the event as some events require special actions from all guests.

That's mostly everything on how to write a perfect main body of an invitation letter. In conclusion, your invitation can contain the following important details:

- Reason for the event.
- Venue.
- The date and day on which the event is taking place.
- Time for arrival.
- List of the special programs and events.

Sample: Invitation letter for an international artificial intelligence conference (See Figure 6. 10)

Figure 6. 10 AI conference

Dear Dr. Richter,

It is my great pleasure to invite you to appear on a panel at the upcoming International Artificial Intelligence Conference. This AI conference will take place on 25 August 2022 in USTC, Hefei, China. We are expecting a packed room, containing some of the most prominent researchers in the field of AI from all over the world, and are eager to end the conference with a panel that summarizes what was new at the conference and points to the future.

The specific topic of the panel is called Machine Learning Reference Architectures. The panelists are Dr. Chen (USTC), Prof. Li (iFlytek Research Institute), Dr. Alice (MIT), Dr. Foster (UOX), and etc. There are over 25 expert speakers and industry experts addressing actual trends and best practices. The event allows individuals to present their research to their audience. I hope you will be willing to discuss the work on creating a reference architecture or a set of standards for ML systems.

Unfortunately, due to budget limitations, we are not able to offer any kind of honorarium or reduced registration fee in return for your appearance on the panel. My sincere regrets.

I do hope you will be able to act as a speaker on the panel. Your experience and comments will add an important dimension to what is potentially a very important discussion for the field.

Sincerely Yours,

David

5.3 Careers in IT

Since artificial intelligence is an increasingly widespread and growing form of technology, jobs in AI have been steadily increasing over the past few years and will continue growing at an accelerating rate. The good news is that the AI professional field is full of different career opportunities. There are some careers in artificial intelligence are shown below.

Data scientist

The main responsibilities of a data scientist are the following:

Selecting features, building and optimizing classifiers using machine learning techniques. Data mining using state-of-the-art methods. Extending company's data with third party sources of information when needed. Enhancing data collection procedures to include information that is relevant for building analytic systems. Processing, cleansing, and verifying the integrity of data used for analysis. Doing ad-hoc analysis and presenting results in a clear manner. Creating automated anomaly detection systems and constant tracking of its performance.

Business intelligence developer

The primary responsibility of a business intelligence developer is to consider the business acumen along with AI. They recognize different business trends by assessing complicated data sets. They help in swelling the profits of a company by preparing, developing and nourishing business intelligence solutions.

Machine learning engineer

Machine learning engineers are involved in building and maintaining self-running software that facilitates machine learning initiatives. They are in continuous demand by the companies and their position rarely remains vacant. They work with huge chunks of data and possess extraordinary data management traits.

They work in the areas of image and speech recognition, prevention of frauds, customer insights, and management of risks. To become a machine learning engineer, one must have sound command in applying predictive models dealing with magnificent data. Programming, computing, and mathematics are essential to becoming successful as a machine learning engineer.

Robotics scientist

A reduction in jobs will indeed take place due to the emergence of robotics in the field of AI. Conversely, jobs will also rise as robotics scientists are in incessant demands by major industries for programming their machines. The robots will help in carrying out certain tasks efficiently.

The candidate should have a master's degree in robotics, computer science or engineering. The median salary for a robotics scientist is quite high. Although automation is favored by robots, there should be some professionals to build them. Thus, the risk of losing jobs is minimized.

6.4 Words and phrases

Artificial Intelligence (AI)
an umbrella term
Computer Vision (CV)
Natural Language Processing (NLP)
speech processing
expert systems
Artificial Narrow Intelligence (ANI)
Artificial General Intelligence (AGI)
Artificial Super Intelligence (ASI)
predict
prevalent
par
surpass
fundamental
illustration
Machine Learning (ML)
imitate
gradually
explicitly
penalty
classification
supervised learning
unsupervised learning
reinforcement learning
Deep Learning (DL)
Artificial Neural Network (ANN)
appropriate
perception
state-of-the-art
venue
convenient
optimizing classifiers
cleanse
anomaly
track
acumen
swell
nourish

specialist	emergence
infrastructure	conversely

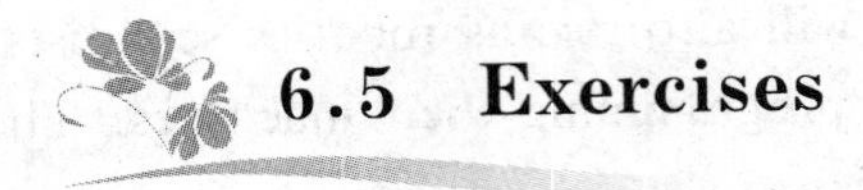

6.5 Exercises

Ⅱ. Matching

Match each numbered item with the most closely related lettered item. Write your answers in the spaces provided.

a. AI
b. machine learning
c. Computer Vision
d. Artificial Neural Networks
e. Deep Learning
f. ASI
g. supervised learning
h. unsupervised learning
i. reinforcement learning
j. Artificial Narrow Intelligence

________(1) a field of Artificial Intelligence (AI) that enables computers and systems to derive meaningful information from digital images, videos and other visual inputs

________(2) an abbreviation for artificial intelligence

________(3) an AI system learns by interacting with its environment

________(4) weak AI

________(5) a term referring to the time when the capability of computers will surpass humans

________(6) an application of AI that provides systems the ability to learn on their own and improve from experiences without being programmed externally

________(7) a subfield of machine learning concerned with algorithms inspired by the structure and function of the brain called artificial neural networks

________(8) an AI system is presented with unlabeled, uncategorized data and the system's algorithms act on the data without prior training

________(9) an AI system is presented with data which is labeled, which means that each data tagged with the correct label

________(10) a term refers to a biologically inspired sub-field of artificial intelligence modeled after the brain

k. Natural Language Processing ________(11) a term refers to the branch of computer science concerned with giving computers the ability to understand text and spoken words in much the same way human beings can

Ⅱ. Written practice

Next month will be your birthday. Please write an invitation letter to friends for your birthday party.

Ⅲ. Open-ended

On a separate sheet of paper, respond to each question or statement.

(1) What's your view about the future of Artificial Intelligence?
(2) How will AI systems enhance and augment human creativity?
(3) Why do we need Artificial Intelligence?

Chapter 7 System and programming

Learning objectives

After you have read this chapter, you should be able to:

☆ Explain system, system analysis, and system design.
☆ Explain stages of system analysis and system design.
☆ Describe the actions of six-phase.
☆ Explain program and programming.
☆ Describe the SDLC.
☆ List the programming languages what you have learned or heard.

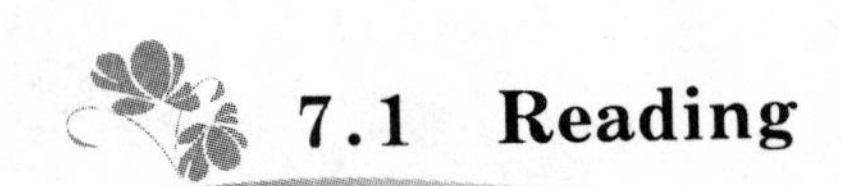

7.1 Reading

Passage 1: System analysis and design

We described different types of information systems. Now let us consider: What, exactly, is a system? We can define it as a collection of activities and elements organized to accomplish a goal. As we saw, an information system is a collection of hardware, software, people, procedures, data, and the Internet. These work together to provide information essential to running an organization. This information helps produce a product or service and, for profit-oriented businesses, derive a profit.

Information about orders received, products shipped, money owed, and so on, flows into an organization from the outside. Information about what supplies have been received, which customers have paid their bills, and so on, also flows within the organization. To avoid confusion, the flow of information must follow a route that is defined by a set of rules and procedures. However, from time to time, organizations need to change their information systems. Reasons include organizational growth, mergers and acquisitions, new marketing opportunities, revisions in governmental regulations, and availability of new technology.

Systems analysis and design is a six-phase problem-solving procedure for examining and improving an information system. The six phases make up the systems life cycle. The phases are as follows (See Figure 7.1):

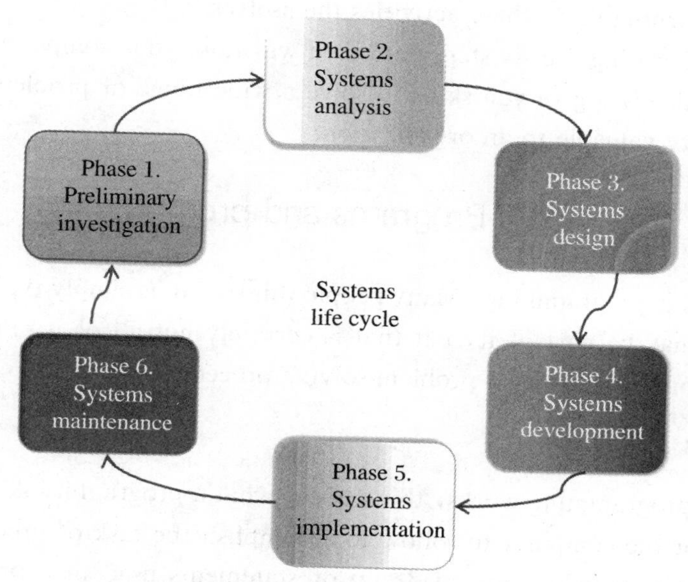

Figure 7.1 System life cycle

(1) **Preliminary investigation**: The organization's problems or needs are identified and summarized in a short report.

(2) **Systems analysis**: The present system is studied in depth. New requirements are specified and documented.

(3) **Systems design**: A new or alternative information system is designed and a design report created.

(4) **Systems development**: New hardware and software are acquired, developed, and tested.

(5) **Systems implementation**: The new information system is installed, and people are trained to use it.

(6) **Systems maintenance**: In this ongoing phase, the system is periodically evaluated and updated as needed.

In organizations, this six-phase systems life cycle is used by computer professionals known as systems analysts. These people study an organization's systems to determine what actions to take and how to use computer technology to assist them.

More and more end users are developing their own information systems. This is because in many organizations there is a three-year backlog of work for systems analysts. For instance, suppose you recognize that there is a need for certain information within your organization. Obtaining this information will require the introduction of new hardware and software. You go to seek expert help from systems analysts in studying these information needs. At that point you discover that the systems analysts are so overworked it will take them three years to get to your request! You can see, then, why many managers are learning to do these activities themselves.

In any case, learning the six steps described will raise your computer efficiency and effectiveness. It also will give you skills to solve a wide range of problems. These skills can make you more valuable to an organization.

Passage 2: Programs and programming

What exactly is programming? Many people think of it as simply typing words into a computer. That may be part of it, but that is certainly not all of it. Programming, as we've hinted before, is actually a problem-solving procedure.

Ⅰ. What is a program?

To see how programming works, think about what a program is. A program is a list of instructions for the computer to follow to accomplish the task of processing data into information. The instructions are made up of statements used in a programming language, such as C++, Java, or Python.

You are already familiar with some types of programs. Application programs are widely used to accomplish a variety of different types of tasks. For example, we use word processors to create documents and spreadsheets to analyze data. These can be purchased and are referred to as prewritten or packaged programs. Programs also can be created or custom-made. Will it do the job, or should it be custom-written? This is one of the first things that needs to be decided in programming.

Ⅱ. What is programming?

A program is a list of instructions for the computer to follow to process data. Programming, also known as software development, typically follows a six-step process known as the Software Development Life Cycle (SDLC) (See Figure 7.2).

The six steps are as follows:

(1) **Program specification**: The program's objectives, outputs, inputs, and processing requirements are determined.

(2) **Program design**: A solution is created using programming techniques such as top-down program design, flowcharts, and logic structures.

(3) **Program code**: The program is written or coded using a programming language.

(4) **Program test**: The program is tested or debugged by looking for syntax and logic

errors.

(5) **Program documentation**: Documentation is an ongoing process throughout the programming process. This phase focuses on formalizing the written description and processes used in the program.

(6) **Program maintenance**: Completed programs are periodically reviewed to evaluate their accuracy, efficiency, standardization, and ease of use. Changes are made to the program's code as needed.

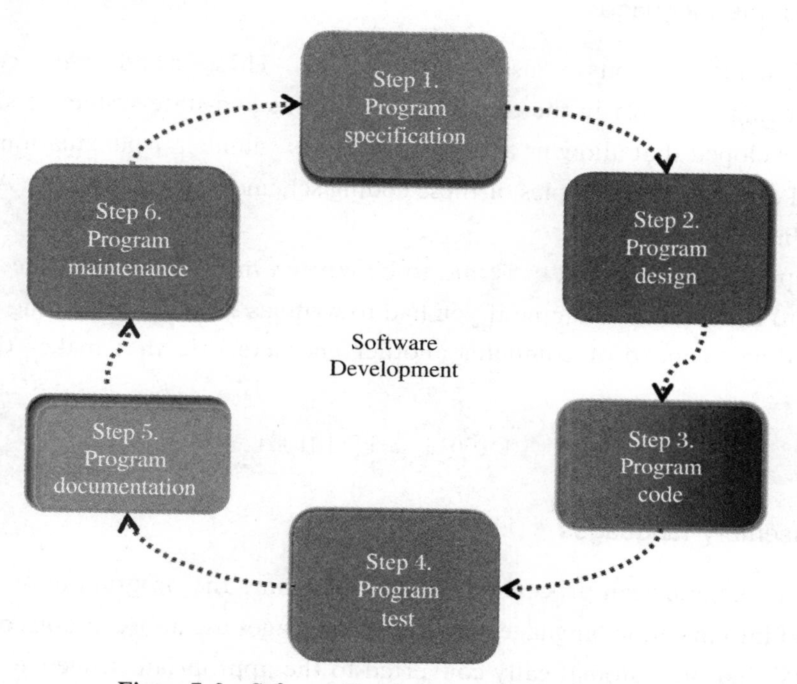

Figure 7.2 Software Development Life Cycle (SDLC)

In organizations, computer professionals known as software engineers or programmers use this six-step procedure. In a recent survey by *Money Magazine*, software engineers were ranked near the top of over 100 widely held jobs based on salary, prestige, and security.

You may well find yourself working directly with a programmer or indirectly through a systems analyst. Or you may actually do the programming for a system that you develop. Whatever the case, it's important that you understand the six-step programming procedure.

Passage 3: Programming languages

Computer professionals talk about levels or generations of programming languages, ranging from "low" to "high". Programming languages are called lower level when they are closer to the language the computer itself uses. The computer understands the 0s and 1s that make up bits and bytes. Programming languages are called higher level when they are closer to the language humans use. That is, for English speakers, more like English.

There are five generations of programming languages: machine languages, assembly languages, procedural languages, task-oriented languages, and problem and constraint languages.

Ⅰ. Machine languages

A byte is made up of bits, consisting of 1s and 0s. These 1s and 0s may correspond to electricity's being on or off in the computer. From this two-state system, coding schemes have been developed that allow us to construct letters, numbers, punctuation marks, and other special characters. Examples of these coding schemes, as we saw, are ASCII, EBCDIC, and Unicode.

Data represented in 1s and 0s is said to be written in machine language. To see how hard this is to understand, imagine if you had to write as shown in following code. It also varies according to make of computer another characteristic that makes them hard to work with.

```
10010011100111101001
```

Ⅱ. Assembly languages

Before a computer can process or run any program, the program must be converted or translated into machine language. Assembly languages use abbreviations or mnemonics such as MOV that are automatically converted to the appropriate sequence of 1s and 0s. Compared to machine languages, assembly languages are much easier for humans to understand and to use. The machine language code we gave above could be expressed in assembly language as shown.

```
ADD AL, 17# add the number 17 and the AL, and result store in the AL.
```

However, the assembly language is also considered low level. Assembly languages also vary from computer to computer. With the third generation, that is, they can be run on more than one kind of computer — they are "portable" from one machine to another.

Ⅲ. High-level procedural languages

People are able to understand languages that are more like English language than machine languages or assembly languages. These more English-like programming languages are called "high-level" languages. Procedural languages, also known as 3GLs (third-generation languages), are designed to express the logic — the procedures that can solve general problems. It intended to solve general problems and are the most widely used languages to create software applications. C++ is a procedural language widely used by today's programmers. For example, C was used as shown:

```
if (max >= 200) display = 'top';
```

Like assembly languages, procedural languages must be translated into machine language so that the computer processes them. Depending on the language, this translation is performed by either a compiler or an interpreter.

Ⅳ. Task-oriented languages

Procedural languages require training in programming. Task-oriented languages, also known as 4GLs (fourth-generation languages) and very high level languages, require little special training on the part of the user as shown in Figure 7.3.

```
1  INSERT INTO Websites (name, url, country)
2  VALUES ('stackoverflow', 'http://stackoverflow.com/', 'IND');
```

Figure 7.3 SQL code

Unlike general-purpose languages, task-oriented languages are designed to solve specific problems. 4GLs are non-procedural and focus on specifying the specific tasks the program is to accomplish. 4GLs are more English-like, easier to program, and widely used by non-programmers. Some of these fourth-generation languages are used for very specific applications.

Query languages enable non-programmers to use certain easily understood commands to search and generate reports from a database. One of the most widely used query languages is SQL (Structured Query Language). For example, let's assume that advantage advertising has a database containing all customer calls for service and that its management would like a listing of all clients who incurred overtime charges.

An application generator or a program coder is a program that provides modules of prewritten code. When using an application generator, a programmer can quickly create a program by referencing the module(s) that performs certain tasks. This greatly reduces the time to create an application. For example, access has a report generation application and a report wizard for creating a variety of different types of reports using database information.

Ⅴ. Problem and constraint languages

As they have evolved through the generations, computer languages have become more human like. Clearly, the fourth-generation query languages using commands that include words like INSERT INTO, and VALUES are much more human-like than the 0s and 1s of machine language. However, 4GLs are still a long way from the natural languages such as English and Spanish that people use.

The next step in programming languages will be the fifth-generation language (5GL), or computer languages that incorporate the concepts of artificial intelligence to allow a person to provide a system with a problem and some constraints, and then request a solution. Additionally, these languages would enable a computer to learn and to apply new information as people do. Rather than coding by keying in specific commands, we would communicate more directly to a computer using natural languages. Consider the following natural language statement that might appear in a 5GL program for recommending medical treatment as shown in Figure 7.4.

When will fifth-generation languages become a reality? That's difficult to say. However, researchers are actively working on the development of 5GL languages and have demonstrated some success.

```
server = smtplib.SMTP('localhost')
server.sendmail('soothsayer@example.org',
    'jcaesar@example.org')
```

Figure 7.4 Natural language statement

Passage 4: Project summary

The original schedule failed in correctly estimating time for design and testing. Therefore, the project team had to spend less time in testing than expected. Luckily, the unit tests were quite efficient, enabling the integration and system tests to be more compact, where the test results were also acceptable. Time estimation in schedule can be more flexible and realistic.

Due to failures of communications facilities, some engineers didn't manage to attend significant meetings. Collaboration among engineers from some different locations can be effectively accomplished electronically via both Internet and telecommunications facilities.

In developing a voice-based application system for local government in China. It would be more effective and more influential for the public to use multi-language voice menus or instructions. The model of multi-language menus provides an option for people with international backgrounds. This feature should be considered for the next release.

This project is intended to develop and implement an Integrated Voice-and Web-Based Public Services System for rural community in China. The accomplishment has fulfilled with the initial project plan, both in the project deliverables and time/budget consideration.

In addition, the main result of this project, voice-based public system, is currently in operation in several Chinese rural communities. The project can be considered as a model in bridging the digital gap in rural China by giving public community access to E-Government public services using Web and voice technologies.

With this Mandarin service, the rural can easily access information and monitor the status of their services provided by government.

All engineers involved in this project obtain some advantages to improve their technical capabilities and large project experiences. Due to failures of communications facilities, some engineers didn't manage to attend certain significant meetings.

In developing a voice-based application system for local government in China, it would be more effective and more influential for the public to use multi-language voice menus or instructions.

The project has been successful in the introduction and implementation of the web-based and voice-based E-Public services in rural area.

This service via installed service machines, web browsers, or telephones. All engineers involved in this project obtain some advantages to improve their technical capabilities and large project experiences and some related human resources of local governments are also trained in operation and maintenance of this web and voice-based E-Public services.

The software applications for public services should be designed with stronger security system because they contain sensitive and secret data. It should ensure that the data be encrypted in the database server using a strong application. Furthermore, an additional function to randomize the registration number and authentication procedure will also be considered in future release.

7.2 Writing: Technical document

Ⅰ. Background overview

1. Introduction

The report study starts with a study of the current environment, the problem within the current environment and a summary of the proposed environment. The general constraints on the development process are summarized in the section that follows:

Operation feasibility study examines how the software will change the roles of the company and customers and whether the workflow and organization structure will be accepted by the customers and company.

Technical feasibility study checks to see if the proposed solution is feasible given the skills of our group and the environment the software is expected to be deployed in the software.

Schedule feasibility study checks if the proposed solution can be developed in a manner that will ensure that all deadlines set by the software engineering and clients are met.

Financial feasibility study examines the costs and benefits of developing the software in the manner of the proposed solution.

2. Scope

This report covers the feasibility of the online ordering system. It does not cover the feasibility of the software is being built for other plug-in components.

Ⅱ. Study of the current and desired environments

The present environment listed as followings:

Technical environment	Name
Programming language	Python
Developing platform	Anaconda & Pycharm
CASE (Computer-aided Software Engineering) tools	Rational Rose
Unit testing tools	testools
Version tools	SVN
Database	Mysql5
DB tools	Mysql Workbench

The desired environment is as followings:

Technical environment	Name
Programming language	Python
Developing platform	PyCharm
CASE tools	Rational Rose
Unit testing tools	test tools
Version tools	none
Database	Mysql
DB tools	Mysql management

Ⅲ. The software requirements

The online ordering system is a way for users to realize online transactions. It has a friendly and intuitive interface and a safe and fast payment method, which makes the seller receive the order information at one time, so that customers can obtain the online ordering information at the first time, and provide customers with higher quality services. It saves time and makes customers feel conveniently and fast. Keep the information of each order properly and handle it in time, realize highly intelligent management, and make ordering faster.

The front desk mainly involves the followings:

(1) **Login and registration**: Login are for old customers, and registration is mainly for new users.

(2) **Books display**: It mainly displays the details of our specialty books, so that users can more intuitively understand and choose their favorite books.

(3) **Front desk information**: Users can browse the books and updating information of the store and evaluating the books.

(4) **Online ordering announcement**: You can release some ordering preferential activities of the store and the specific time of the discount.

(5) **Query module**: Users can query their favorite books, and then learn about the experience of other users in the comment area.

(6) **Contact us**: It is mainly used to feedback relevant suggestions on some dishes and other services, let us know the shortcomings, continuously improve and publicize them at the same time and users can contact at any time.

Ⅳ. Constraints

Technical constraint is that the system has to work on the windows and Android and iOS. The other technical constraint is a result the skill of the team. Most of them are familiar with Java and t Python but C+ + (financial and schedule constraints omitted).

Ⅴ. Possible solution

Solution

Operation feasibility

(1) System security

The online ordering system should control the use rights of different users. Users should not operate beyond their authority. The system should ensure the security of data. At the same time, the system is required to have high reliability, data accuracy and system recoverability.

(2) Maintainability

Customers will constantly put forward new requirements for the system and expand the system functions during the use of the system, which requires that the system must be well upgradeable to meet the requirements of customers for long-term use, and can expand rapidly when users have new requirements.

(3) User operation

The system design should be humanized, easy to operate and interface friendly.

(4) Running speed

Response time of the system is an important reference to measure system performance. The system requires short response time, rapid update processing, short conversion and transmission time. Due to the particularity of system users, the system must respond efficiently and respond efficiently on the premise of safety.

(5) Interface

The user interface should be clear, simple, clear at a glance and easy to operate.

(6) Technical feasibility

Online ordering system needs database technology, network technology and related development technology. These technologies are mature at present, and the development of this system is completely feasible.

(7) Schedule feasibility

We can meet all deadline established throughout the development process because the project is not complicated and big.

(8) Financial feasibility

The investment in developing this system is small. The enterprise has the ability to bear the development and daily maintenance costs of the system. In the later stage, it can invest in advertising for profit, save a lot of material and human resources in ordering meals, and have good economic benefits.

(9) Legal feasibility

The preparation before system development and the whole process do not violate laws and regulations, and do not conflict with the current management system.

Ⅵ. Conclusion

By examining the result of the feasibility study, the project team has decided the first solution is the most feasible. It is also to represent a more stable and reliable system.

Above-mentioned document. Keep these questions in mind and try to fill the following Table 7.1 with them after your scan.

Table 7.1 Project report sample

Project Report			
Background overview	Introduction	Content 1	Content 2
	Scope		
Study of the current and desired environment	Present environment		
	Problems with present environment		
	Desired environment		
Software requirements	Requirements acquisition		
	Requirements analysis		
	Requirements specifications		
	Requirements review		
Constraints	Technical constraints		
	Financial constraints		
	Schedule constraint		
Option solutions	Solution	Operation feasibility	
		Technical feasibility	
		Schedule feasibility	
		Financial feasibility	
		Legal feasibility	
Conclusion	Summarizing project		

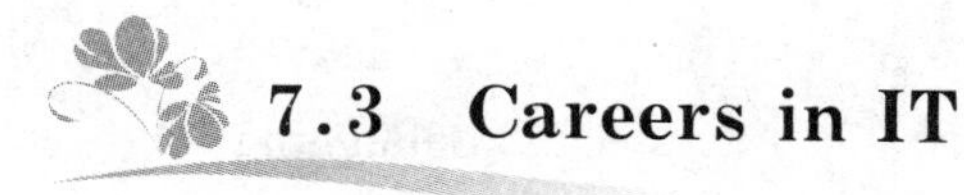

7.3 Careers in IT

A systems analyst

Primary duties: A systems analyst reviews design components and uses their knowledge of information technology to solve business problems. They identify ways that infrastructure needs to change to streamline business and IT operations. They can also assist technicians in training staff to implement the changes they propose.

Requirements: A Bachelor's Degree in Computer Science or a related field is often required. Coursework in business administration, management and finance may help these professionals better apply their IT knowledge to improving business practices.

Software engineers

Software engineers analyze users' needs and create application software. Software engineers typically have experience in programming but focus on the design and development of programs using the principles of mathematics and engineering.

A bachelor's or an advanced specialized associate's degree in computer science or information systems and an extensive knowledge of computers and technology are required by most employers. Internships may provide students with the kinds of experience employers look for in a software engineer. Those with specific experience with web applications may have an advantage over other applicants. Employers typically look for software engineers with good communication and analytical skills.

Software engineers can expect to earn an annual salary in the range of ￥100, 000 to ￥250, 000. Starting salary is dependent on both experience and the type of software being developed. Experienced software engineers are candidates for many other advanced careers in IT.

7.4 Words and phrases

agile development
application generator
assembly language
beta testing
code
code review
coding
compiler
Computer-aided Software Engineering (CASE) tools
debugging
desk checking
documentation
fifth-generation language (5GL)
fourth-generation language (4GL)
generation
higher level
IF-THEN-ELSE structure
Interactive Financial Planning System (IFPS)

interpreter
level
logic error
logic structure
loop structure
lower level
machine language
maintenance programmer
module
natural language
object
object code
object-oriented programming (OOP)
object-oriented software
development
objective
operator
patches
portable language
procedural language
program
program analysis
program coder
program definition
program design
program documentation
program flowchart
program maintenance
program module
program specification
programmer
programming
programming language
query language
repetition structure
selection structure
sequential structure
software development
Software Development Life Cycle (SDLC)
software engineer
software updates
source code
structured program
structured programming technique
syntax error
task-oriented language
third-generation language (3GL)
top-down program design
user
very high level language

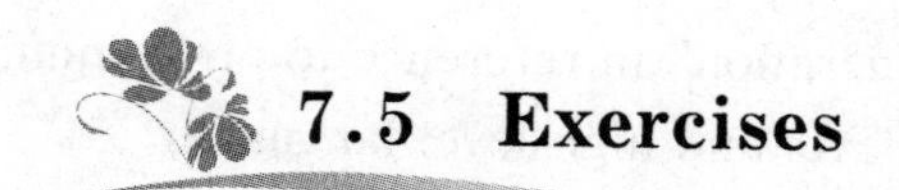

7.5 Exercises

Ⅰ. Matching

Match each numbered item with the most closely related lettered item. Write your answers in the spaces provided.

a. debugging ________(1) six-step procedure also known as software development

b. documentation ________(2) an outline of the logic of the program to be written

c. higher level ________(3) logic structure, also known as IF-THEN-ELSE, that controls program flow based on a decision

d. interpreter ________(4) programming languages that are closer to the language of humans

e. machine ________(5) the process of testing and then eliminating program errors

f. natural language ________(6) program step that involves creating descriptions and procedures about a program and how to use it

g. programming ________(7) the first-generation language consisting of 1s and 0s

h. pseudocode ________(8) converts a procedural language one statement at a time into machine code just before it is to be executed

i. selection ________(9) generation of computer languages that allows a person to provide a system with a problem and some constraints, and then request a solution

j. 5GL ________(10) 5GL that allows more direct human communication with a program

Ⅱ. Written practice

Write a project report about your IT English training. Write this report both as a summary of your work during the training and as a demonstration of writing final reports that comply with high standards of communication as to their organization, format, content, and style.

Ⅲ. Open-ended

On a separate sheet of paper, respond to each question or statement.

(1) Identify and discuss each of the six steps of programming.

(2) Describe CASE tools and OOP. How does CASE assist programmers?

(3) What is meant by "generation" in reference to programming languages? What is the difference between low-level and high-level languages?

Chapter 8 Cloud computing

Learning objectives

After you have read this chapter, you should be able to:

☆ Understand the definition and characteristic of cloud computing.
☆ Compare the cloud computing and traditional private.
☆ Understand the benefit of cloud computing.
☆ Compare each cloud service mode.
☆ Describe the different level of responsibility that cloud provider and cloud tenant are responsible for.

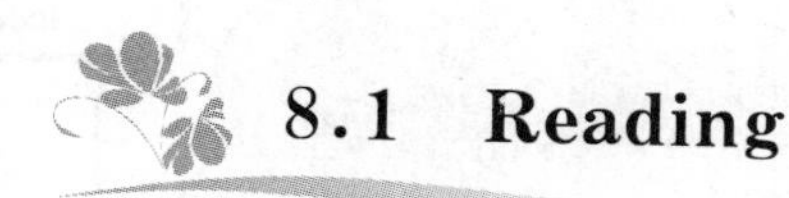

8.1 Reading

Passage 1: Cloud computing

Ⅰ. Definition of cloud computing

Have you ever wondered what cloud computing is? It's the delivery of computing services via the internet, which is known as the cloud. These services include servers, storage, databases, networking, software, analysis, and intelligence. Cloud computing offers faster innovation, flexible resources, and economies of scale (See Figure 8.1).

The National Institute of Standards and Technology's definition of cloud computing identifies "five essential characteristics" (See Figure 8.2):

1. On-demand self-service

A consumer can unilaterally provision computing capabilities, such as server time and network storage, as needed automatically without requiring human interaction with each service provider.

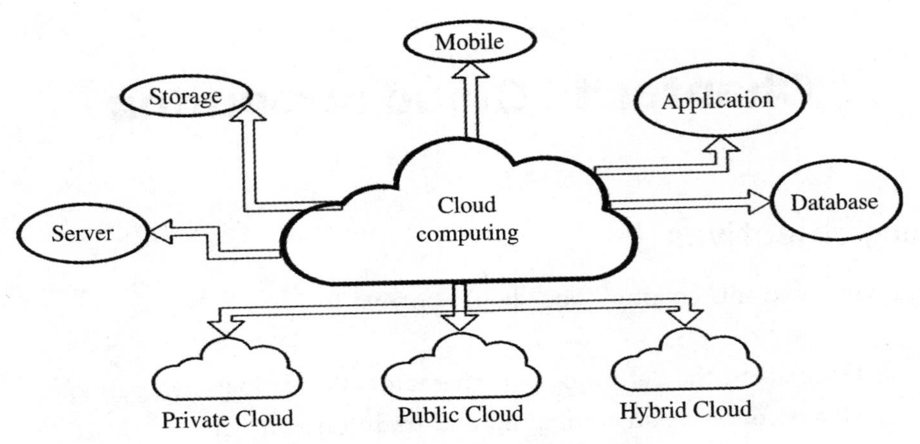

Figure 8.1 Cloud computing outline

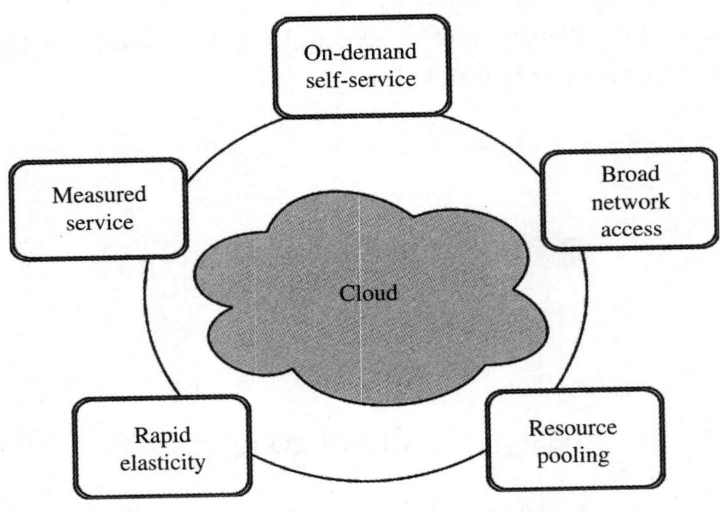

Figure 8.2 Characteristic of cloud computing

2. Broad network access

Capabilities are available over the network and accessed through standard mechanisms that promote use by heterogeneous thin or thick client platforms (e.g., mobile phones, tablets, laptops, and workstations). Cloud computing service models arranged as layers in a stack.

3. Resource pooling

The provider's computing resources are pooled to serve multiple consumers using a multi-tenant model, with different physical and virtual resources dynamically assigned and reassigned according to consumer demand.

4. Rapid elasticity

Capabilities can be elastically provisioned and released, in some cases automatically, to scale rapidly outward and inward commensurate with demand. To the consumer, the capabilities available for provisioning often appear unlimited and can be appropriated in any quantity at any time.

5. Measured service

Cloud systems automatically control and optimize resource use by leveraging a metering capability at some level of abstraction appropriate to the type of service (e.g., storage, processing, bandwidth, and active user accounts). Resource usage can be monitored, controlled, and reported, providing transparency for both the provider and consumer of the utilized service.

Ⅱ. What are the benefits of cloud computing?

There are several benefits that a cloud environment has over a physical environment. For example, cloud-based applications employ a myriad of related strategies:

1. Reliability

Depending on the service-level agreement that you choose, your cloud-based applications can provide a continuous user experience with no apparent downtime even when things go wrong.

2. Scalability

Applications in the cloud can be scaled in two ways, while taking advantage of autoscaling:

Vertically: Computing capacity can be increased by adding RAM or CPUs to a virtual machine.

Horizontally: Computing capacity can be increased by adding instances of a resource, such as adding more virtual machines to your configuration.

3. Elasticity

Cloud-based applications can be configured to always have the resources they need.

4. Agility

Cloud-based resources can be deployed and configured quickly as your application requirements change.

5. Geo-distribution

Applications and data can be deployed to regional data-centers around the globe, so your customers always have the best performance in their regions.

6. Disaster recovery

By taking advantage of cloud-based backup services, data replication, and Geo-distribution, you can deploy your applications with the confidence that comes from knowing that your data is safe in the event that disaster should occur.

Passage 2: Cloud service models

Ⅰ. What are cloud service models?

Cloud computing falls into one of the following computing models. If you've been around cloud computing for a while, you've probably seen the terms infrastructure as a service (IaaS), platform as a service (PaaS), and software as a service (SaaS) for the different cloud service models. These models define the different level of shared responsibility that a cloud provider and cloud tenant are responsible for (See Table 8.1).

Table 8.1 Cloud service models

Models	Description
IaaS	This cloud service model is the closest to managing physical servers. A cloud provider keeps the hardware up to date, but operating system maintenance and network configuration is left to the cloud tenant. For example, Azure virtual machines are fully operational virtual compute devices running in Microsoft's datacenters. An advantage of this cloud service model is rapid deployment of new compute devices. Setting up a new virtual machine is considerably faster than procuring, installing, and configuring a physical server.
PaaS	This cloud service model is a managed hosting environment. The cloud provider manages the virtual machines and networking resources, and the cloud tenant deploys their applications into the managed hosting environment. For example, Azure App Services provide a managed hosting environment where developers can upload their web applications without having to deal with the physical hardware and software requirements.
SaaS	In this cloud service model, the cloud provider manages all aspects of the application environment, such as virtual machines, networking resources, data storage, and applications. The cloud tenant only needs to provide their data to the application managed by the cloud provider. For example, Office 365 provides a fully working version of Office that runs in the cloud. All that you need to do is to create your content, and Office 365 takes care of everything else.

Figure 8.3 illustrates the various levels of responsibility between a cloud provider and a cloud tenant.

Table 8.2 illustrates the various levels of responsibility between a cloud provider and a cloud tenant.

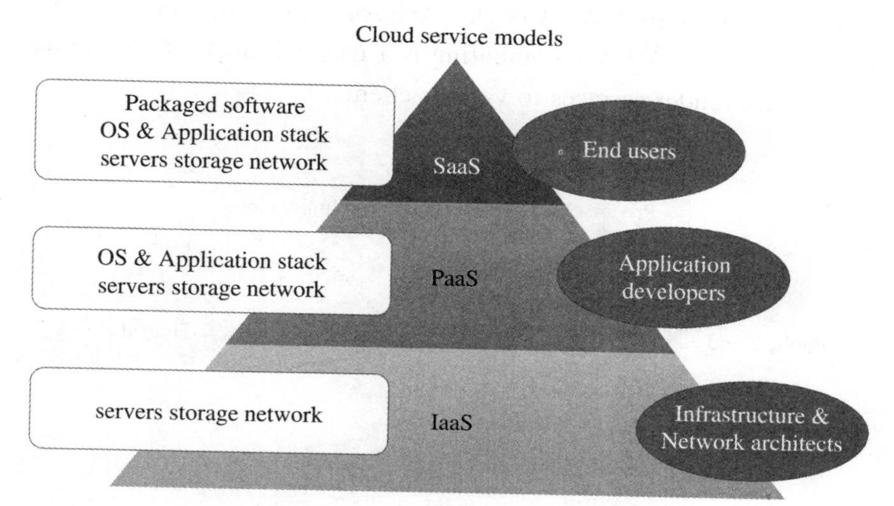

Figure 8.3 Cloud service models

Table 8.2 Responsibility between provider and tenant

	Private cloud	IaaS	PaaS	SaaS
Storage				
Networking				
Compute				
Virtual machine				
Operating system				
Runtime				
Application				
Data & Access				

You Manage: ______________________________

Cloud Provider: ______________________________

Ⅱ. What are the cloud computing deployment model?

There are four types of them: public, private, hybrid and community clouds. Cloud deployment model represents the exact category of cloud environment based on proprietorship, size, and access and also describes the nature and purpose of the cloud (See Figure 8.4). Most organizations implement cloud infrastructure to minimize capital expenditure and regulate operating costs.

1. Public cloud

Public cloud is a type of cloud hosting that allows the accessibility of systems & its services to its clients/users easily. Some of the examples of those companies which provide public cloud facilities are IBM, Google, Amazon, Microsoft, etc. This cloud service is open for use. This type of cloud computing is a true specimen of cloud hosting where the service providers render services to various clients.

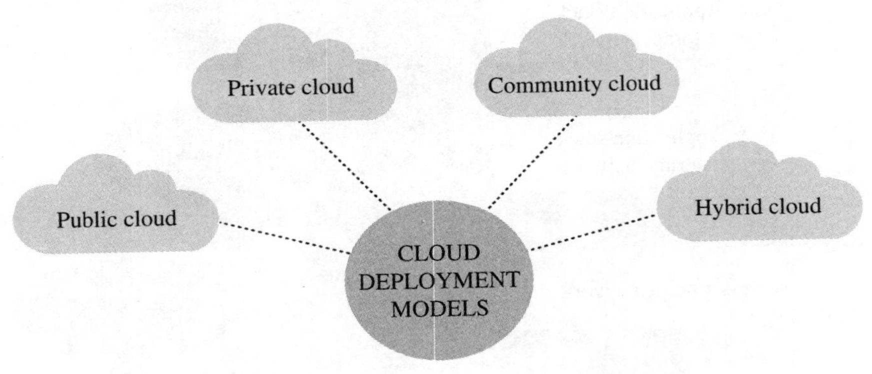

Figure 8.4 Cloud deployment models

Server infrastructure belongs to service providers that manage them and administer pool resources, which is why there is no need for user companies to buy and maintain their hardware. Provider companies offer resources as a service both free of charge or on a pay-per-use basis via the Internet connection. Users can scale resources when required.

The public cloud deployment model is the first choice for businesses that operate within the industries with low privacy concerns. When it comes to popular public cloud deployment models, examples are Amazon Elastic Compute Cloud (Amazon EC2) the top service provider, Microsoft Azure, Google App Engine, IBM Cloud, Salesforce Heroku and others.

2. Private cloud

There is little to no difference between a public and a private model from the technical point of view, as their architectures are very similar. However, opposed to a public cloud that is available to the general public, only one specific company owns a private one. That is why it is also called an internal or corporate cloud.

"Internal cloud" means that it allows the accessibility of systems and services within a specific boundary or organization. The cloud platform is implemented in a cloud-based secure environment that is guarded by advanced firewalls under the surveillance of the IT department that belongs to a particular organization. Private clouds permit only authorized users, providing the organizations greater control over data and its security. Business organizations that have dynamic, critical, secured, management demand-based requirement should adopt private cloud.

3. Community cloud

A community deployment model largely resembles a private one; the only difference is the set of users. While a private type implies that only one company owns the server, in the case of a community one, several organizations with similar backgrounds share the infrastructure and related resources. Example of such a community is where organizations/firms are there along with the financial institutions/banks. A multi-tenant setup developed using cloud among different organizations that belong to a particular community or group having similar computing concern.

For joint business organizations, ventures, research organizations and tenders community cloud is the appropriate solution. Selection of the right type of cloud hosting is essential in this case. Thus, community-based cloud users need to know and analyze the business demand first.

4. Hybrid Cloud

Hybrid Cloud is another cloud computing type, which is integrated, i.e., it can be a combination of two or more cloud servers, i.e., private, public or community combined as one architecture, but remain individual entities. Non-critical tasks such as development and test workloads can be done using public cloud whereas critical tasks that are sensitive such as organization data handling are done using a private cloud. Benefits of both deployment models, as well as a community deployment model, are possible in a hybrid cloud hosting.

Table 8.3 The advantages and disadvantages of deployment models

Model	Advantages	Disadvantages
Public cloud	• Flexible • Reliable • High scalable • Low cost • Place independence	• Less secured • Poor customizable
Private cloud	• Highly private and secured: Private cloud resource sharing is highly secured. • Control oriented: Private clouds provide more control over its resources than public cloud as it can be accessed within the organization's boundary.	• Poor scalability: Private type of clouds is scaled within internal limited hosted resources. • Costly: As it provides secured and more features, so it's more expensive than a public cloud. • Pricing: is inflexible; i.e., purchasing new hardware for up-gradation is more costly. • Restriction: It can be accessed locally within an organization and is difficult to expose globally.

Continued

Model	Advantages	Disadvantages
Community cloud	• Cost reduction • Improved security, privacy and reliability • Ease of data sharing and collaboration	• High cost if compared to a public deployment model. • Sharing of fixed storage and bandwidth capacity. • It is not widespread so far.
Hybrid cloud	• Flexible • Secure • Cost effective • Rich scalable	• Complex networking problem. • Organization's security Compliance.

Hybrid cloud deployment model not only safeguards and controls strategically important assets but does so in the most cost-saving and resource-effective way possible for each specific case. Also, this approach facilitates data and application portability.

8.2 Writing: How to write an email for requesting

There are many times when we need something. It can be leave from your boss, an interview at a new company, business advice from an expert, or even a recommendation. It can be daunting to write request emails that will get the recipient to grant your request. In this article, we will show you how to write an email requesting something and actually get a response!

Ⅰ. What is a request email?

As the name suggests, a request email is an email you write, asking for something, whether information, favor, or service. The email can be to ask for help, authorization, advice, support, etc. It can also be an appeal or inquiry. Since it's a request, the email should be very polite, precise, brief, and specific. The recipient should be able to know what you want by the end of the email.

Ⅱ. Preparation for writing a request email

1. Focus on your perspective

When requesting something from the recipient, it can be tempting to sink into talking about yourself. The secret is to make it about the recipient. Tell them how you appreciate what they have been doing. Let them know you are a big fan of their work and how their work has transformed your life before asking for an interview. The "you first" approach will help you get backstage passes, several leaves a year, interviews and recommendation letters with minimal effort. Remember, focus on them!

2. Sell your benefits

What value are you adding to the recipient so that you should secure an interview? If you ask for a favor, an internship, or just an interview, ensure you have done thorough research on the pain points and how you can provide solutions. Set yourself apart!

3. Make it impossible to say no

How? Anticipate rejection and come up with a solution to their reasons for saying no. For instance, if you want to be an intern for the best-selling writer, they may say no because they don't have a budget for your salary. Offer to intern for free. You will be making it impossible to say no.

Ⅲ. Structure

The structure of a request email is similar to that of an official letter. Keep it formal and respectful to increase the chances of the person doing what you are asking.

1. Subject line

The subject line should state why you are emailing the person. It will determine whether the recipient opens your email or not. Keep it short but precise. For instance, "requesting a recommendation letter."

2. Salutation/email introduction

The salutation should be formal unless you know the recipient personally. If you know the name of the recipient, use it to create a personal feeling. For instance, "hello Josh" is likely to make the recipient open the email than "hello sir."

3. Body

The body of a request is very simple to craft. Just remember the acronym rap, which stands for reference, action, and polite close. In reference, let the recipient know why you are writing. For instance, "I am writing to request a recommendation for my internship at [**company**]." The recipient does not have to read through the whole email to know what you want. Ensure you are using polite language. You can start off with phrases like:

I am writing to request ...

With reference to ...

I am writing in response to your inquiry ...

Under action, clearly state what you want the recipient to do for you. For instance, "please send the documents before evening for compiling." The action should also be very polite because it's a request. Remember, you are not entitled to what you are requesting. Also, keep it brief and straight to the point. A long body is likely to discourage the recipient from reading the email.

Finish by thanking the recipient for the time spent reading the email. You can use polite phrases like:

Thanks in advance for your assistance.

Please let me know if you have any questions.

I look forward to hearing from you.

Please let me know when you are available.

4. Email ending

The email ending also creates an impression, so ensure you keep it as professional and precise as the rest of the email. Keep it simple, for instance:

Best regards,

[**name**]

Ⅳ. Email do's and don'ts

1. Email do's

(1) Be precise. Nobody wants to read a novel, especially when you are writing to them, taking up their time and requesting something from them. No need for beating around the bush. Keep it short and precise!

(2) Limit your email to one request. You may want a thousand things, but don't confuse the recipient. You will end up burying the most important request or even getting nothing out of it!

(3) Use a polite tone and language. A request email should be very polite, like someone who is asking for something and not demanding. Remember that the recipient owes you nothing. Once you are done writing, try reading it from the recipient's perspective. Would you grant a request to the writer based on the tone?

(4) Proofread your email. You don't want your recipient to be put off by poor grammar or punctuation. Ensure you have proofread your email. You can make sure that your email is clear and free of mistakes by using a writing tool such as Grammarly. This will greatly increase your chances of success when making your request.

2. Email don'ts

(1) Don't write anything inappropriate in the email. Remember, emails are not private.

(2) Don't forget to proofread. It does not matter if you are in a hurry. Poor grammar is a turn-off.

(3) Don't send an email if it can be addressed in person or over the call. People have too many emails every day. They don't want more!

Ⅴ. Email sample

Are you still having trouble writing an email to ask for something? We got you! Here are some samples you can customize to suit your needs.

Sample: Leave request email

Subject line: Request for one week leave

Dear [**name**]

With reference to our meeting in the afternoon, I would like to request one week leave. I have been following up with the interns, and I am feeling a bit under the weather due to working late.

I have been monitoring my health for 10 days, but there has been no improvement, I would like to take time off to see a doctor and get back on my feet to do a better job. I have ensured that all tasks I cannot handle online will be attended to by [**name of colleague**] so everything can run smoothly in the department.

I will be awaiting your response. Thanks in advance.

Sincerely,

[**name**]

8.3 Careers in IT

Cloud computing touches many aspects of modern life, and the need for cloud professionals is great.

Cloud computing engineer

This position will help build and monitor highly resilient, secure, and cost-effective cloud computing environments that supports and enriches research productivity and reliability following best practices standards. This position is customer-focused and will work closely with researchers for gathering requirements, systems architects for creating cost estimates, and with other systems teams members for setting up Windows/Linux environments and supporting applications for research projects.

Job responsibilities include:

• Technical Consultation: Work closely with researchers to help set up research project environments in the cloud, including provisioning and de-provisioning, deploying, managing, operating cloud resources, and supporting web applications. Begin to build an understanding of research activities through regular engagements. Assist in creating cost estimates for cloud projects.

• Software Development: Supports UCF Research Cloud Environment (developing, deploying, releasing services and applications in the cloud, debugging, and troubleshooting issues when necessary). Write scripts when needed. Keep abreast of security best practices. Follow a scope of work, project plan, working on and track the progress of small number of milestones at a time.

• Partnership/Collaboration: Work collaboratively with other team engineers and researchers to build and maintain the applications and services that sit on top of the cloud infrastructure. Provide regular communications to project leads with updates.

• Documentation/Training: Maintain internal guides for future contributors. Assist in creating training material for researchers about usage and best practices of cloud environments. Grow skills in a specific technology to develop custom solutions to meet researcher needs.

• Other duties as assigned.

8.4 Words and phrases

accessibility
agility
analysis
architecture
automatically
boundary
capability
collaboration
community cloud
compliance
configuration
considerably
consumer demand
corporate
customizable
delivery
deployment
disaster recovery
elasticity
essential
expenditure
flexible

Geo-distribution
horizontally
hybrid cloud
IaaS
illustrate
innovation
leverage
maintenance
measured service
mechanisms
PaaS
perspective
physical
private cloud
proprietorship
provision
public cloud
rapid elasticity
regional
reliability
resource pooling
responsibility
restriction
SaaS
specimen
storage
surveillance
tenant
transparency
unilaterally
utilize

8.5 Exercises

Ⅱ. Matching

Match each numbered item with the most closely related lettered item. Write your answers in the spaces provided.

a. scalability	________(1) the basic cloud service model which cloud provider keeps physical servers, network and other hardware
b. Geo-distribution	________(2) application can get enough resources they need in the could
c. elasticity	________(3) used for one business or organization, and this can be physically located in their own datacenter
d. IaaS	________(4) applications and data can be deployed to regional da-ta-center around the world
e. PaaS	________(5) applications can be scaled in two ways while in the cloud
f. SaaS	________(6) services are offered over the public internet and available to anyone who wants to purchase them

g. public cloud ________(7) the middle cloud service model which cloud provider keeps develop environment, deploy environment, such as Azure App Services

h. private cloud ________(8) combines a public cloud and a private cloud by allowing data and applications to be shared between them

i. hybrid cloud ________(9) cloud provider manages all aspects of application environment, cloud tenant only needs to provide their data, like Office 365

Ⅱ. Written practice

Letter of recommendation request email

Subject: Requesting a recommendation letter

Your Professor: David

A good job you want get it of HUAWEI, so you want your professor help you to write a recommendation letter. Then you can go ...

Ⅲ. Open-ended

On a separate sheet of paper, respond to each question or statement.

(1) What is cloud computing? Describe three basic components of cloud computing.

(2) What use cases can you imagine for cloud computing in the future?

(3) What are the advantages and disadvantages of using cloud computing?

Chapter 9 Privacy, security and ethics

Learning objectives

After you have read this chapter, you should be able to:

☆ Explain what is cyber privacy and personal data.
☆ Get to know what is computer security and ethics.
☆ Be aware of the importance to protect personal data computer security.
☆ Be aware of the importance to abide by computer ethics.
☆ Be aware of the various cybercrimes and attacks.
☆ Get to know effective measures to prevent computers and networks from hacking, attacking and phishing.
☆ Be familiar with related words and vocabularies.

In this digital era, computers and IT technology are pervasive. The Internet and the web connect individual computers and people together, and it's very probably for our computers and personal data to be hacked and attacked. Technology is a sword with two sides, you need to stay cautious about the security of your personal computer and data. It's extremely important to learn how to protect your computer system and personal data from hacking, attacking and various cybercrimes. Meanwhile, although computers seem can do everything, there are certain computer ethics to be followed when they are used in our society, determining what you can't do with computers.

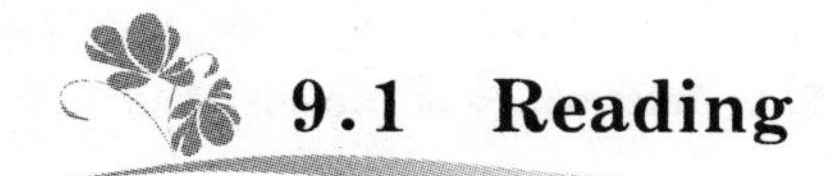

9.1 Reading

Passage 1: Protect your cyber privacy

The term "privacy" is very common in daily life. At this information age, however, it mainly concerns the collection and use of data about individuals, which deserves extremely cautious treatment by the individual concerned.

It's not exaggeration! Nowadays, more than one billion personal computers are in use every day. Does the widespread presence of this technology bear hazard for computer users? Will this technology make it easy for others to invade our personal privacy? For example, when we apply for a credit card or for a driver's license, or when we shop online, is there possibilities for our personal information being distributed and used without our permission? When we logon a website, is information about us being secretly collected and shared with others?

Unfortunately, answers to the above questions are positive. This technology prompts a lot of questions, among which these are some of the most important ones for the 21st century. To use computers more efficiently and effectively, you need to be conscious of the potential impact of technology on people, meanwhile, you are supposed to be aware of measures to protect yourself on the web. To guarantee your personal privacy and security in cyberspace, you have to be sensitive to and knowledgeable about these questions and corresponding measures (See Figure 9.1).

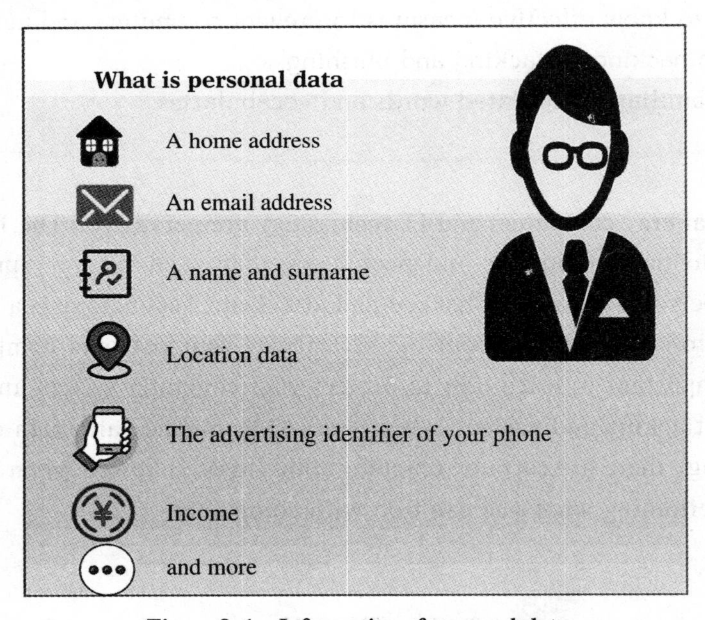

Figure 9.1 Information of personal data

The issue of privacy becomes increasingly critical at an age since millions of personal data may be easily collected and sold by information brokers (See Figure 9.2). Then through analysis of the information a colossal volume of electronic profiles is likely to be compiled by large organizations from these large databases, or referred as "big data", providing highly detailed and personalized descriptions of any individual user. This may result in serious consequences on economic, social or even physical safety problems of any targeted individuals. For instance, mistaken identity occurs when we log on a registered account of a website, which probably an indication of your identity having been switched with another. Besides, many organizations monitor emails and files of their employees via special software. Therefore, unless carefully protected, personal privacy is vulnerable to outside intrusion on the web.

Figure 9.2 Keep computer privacy in mind

Although there are still many people believing little can be done to invade their privacy, it's definitely an illusion!

As a result, you may be concerned about how to protect your privacy while using the web. The following are some suggestions for your reference.

(1) Using privacy mode while browsing the Internet. Since browsers store information such as history files and browser cache, privacy mode is suggested for you as it ensures your browsing activity is not recorded.

(2) Prohibit cookies when browsing websites containing sensitive personal information. Cookies also can store and track information of users. On one hand, they help you browse the web more conveniently; on the other hand, it exposes your personal data. When the website you are browsing involves your privacy information, prohibit the cookies from operation is advisable.

(3) Install anti-spyware to remove spy programs. It is very possible that the software you download and install turns out to be a spyware which secretly records and reports your activities on Internet. For example, the keystroke loggers, a type of computer monitoring software, records every activity and keystroke. Anti-spyware helps you detect and remove various privacy threats from the cyber space.

(4) Think twice before posting personal information and intimate details of private lives on Internet.

The personal information and details of your private life creates an online identity, which is available to anyone aiming to take advantage of it and might even fall into hands of criminals. As a result, please consider the consequences before posting privacy information of yourself.

In conclusion, since personal privacy seems so susceptible to intrusion, it's compelling for every user to bear personal privacy in mind and take effective measures to protect it.

Passage 2: Computer security for individuals and organizations

As various risks posing against computer users, the issue of computer security is inevitably a common concern at the digital age (See Figure 9. 3). Computer security focuses on how to "protect information, hardware, and software from unauthorized use as well as preventing damage from intrusions, sabotage, and natural disasters". Speaking of unauthorized access to people's information on Internet, we usually think it's a hacker's practice. Actually, however, not all hackers intend to maliciously attack computers, and sometimes they are not necessarily criminals.

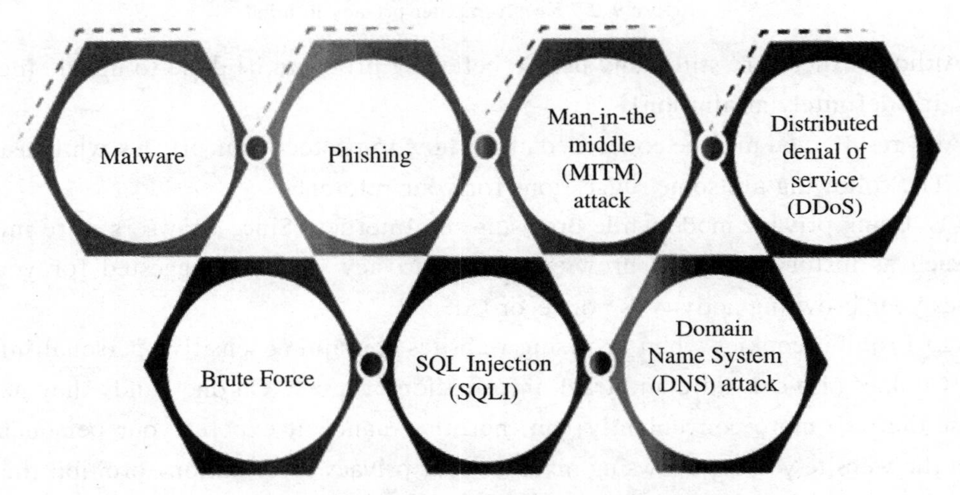

Figure 9. 3 Types of computer security risks

The first step to protect computer security for individuals and organizations is to get acquainted with various cybercrimes, which are illegal actions through manipulating computer skills to aim at stealing, deceiving or destroying computer information, financial property and malicious attack against computers or people. Here are some cybercrime examples demanding our attention:

1. Malware

Malware means malicious software, which is the most common cyber attacking tool. It is used by the cybercriminal or hacker to disrupt or damage a legitimate user's system. The following are the important types of malware created by the hacker:

- **Virus**: It is a malicious piece of code that spreads from one device to another. It can clean files and spreads throughout a computer system, infecting files, stoles information, or damage device.

• **Spyware**: It is a software that secretly records information about user activities on their system. For example, spyware could capture credit card details that can be used by the cybercriminals for unauthorized shopping, money withdrawing, etc.

• **Trojans**: It is a type of malware or code that appears as legitimate software or file to fool us into downloading and running. Its primary purpose is to corrupt or steal data from our device or do other harmful activities on our network.

• **Ransomware**: It's a piece of software that encrypts a user's files and data on a device, rendering them unusable or erasing. Then, a monetary ransom is demanded by malicious actors for decryption.

• **Worms**: It is a piece of software that spreads copies of itself from device to device without human interaction. It does not require them to attach themselves to any program to steal or damage the data.

• **Adware**: It is an advertising software used to spread malware and displays advertisements on our device. It is an unwanted program that is installed without the user's permission. The main objective of this program is to generate revenue for its developer by showing the ads on their browser.

• **Botnets**: It is a collection of internet-connected malware-infected devices that allow cybercriminals to control them. It enables cybercriminals to get credentials leaks, unauthorized access, and data theft without the user's permission.

2. Denial of Service (DoS) attack

This is a cyberattack on computer devices, information systems, or network resources. It prevents legitimate users from accessing network services and resources. By flooding the targeted host computer or network with tremendous meaningless requests, this attack can block legitimate access and request, and eventually lead to crashes of the host or network. It may cause a huge loss to organizations for failure in obtaining resources and services.

3. Phishing

Phishing is a type of cybercrime in which a sender seems to come from a genuine organization like PayPal, eBay, financial institutions, or friends and co-workers. They contact a target or targets via email, phone, or text message with a link to persuade them to click on that links. This link will redirect them to fraudulent websites to provide sensitive data such as personal information, banking and credit card information, social security numbers, usernames, and passwords. Clicking on the link will also install malware on the target devices that allow hackers to control devices remotely.

4. Man-in-the-middle (MITM) attack

A man-in-the-middle attack is a type of cyber threat (a form of eavesdropping attack) in which a cybercriminal intercepts a conversation or data transfer between two individuals. Once the cybercriminal places themselves in the middle of a two-party communication, they seem like genuine participants and can get sensitive information and return different responses. The main objective of this type of attack is to gain access to our business or customer data. For example, a cybercriminal could intercept data passing between the target device and the network on an unprotected Wi-Fi network.

5. Brute force

A brute force attack is a cryptographic hack that uses a trial-and-error method to guess all possible combinations until the correct information is discovered. Cybercriminals usually use this attack to obtain personal information about targeted passwords, login info, encryption keys, and Personal Identification Numbers (PINS).

6. SQL Injection (SQLI)

SQL Injection is a common attack that occurs when cybercriminals use malicious SQL scripts for backend database manipulation to access sensitive information. Once the attack is successful, the malicious actor can view, change, or delete sensitive company data, user lists, or private customer details stored in the SQL database.

7. Domain Name System (DNS) attack

A DNS attack is a type of cyberattack in which cyber criminals take advantage of flaws in the Domain Name System to redirect site users to malicious websites (DNS hijacking) and steal data from affected computers. It is a severe cybersecurity risk because the DNS system is an essential element of the internet infrastructure.

There are many ways to protect computer security. First, restrict access through biometric scanning devices and passwords.

Second, Encrypting is effective in coding information to make it unreadable to those who don't have the encryption key. Currently, some forms of encryption that can effectively prevent data from outside invasion are available. To name a few, Hypertext Transfer Protocol Secure (HTTPS), Virtual Private Networks (VPNs), and Wi-Fi Protected Access (WPA2) are some most widely used encryptions.

Lastly, to prevent data loss employers need to screen job applicants in recruitment, be careful to guard passwords (usually by setting strong passwords) and back up data frequently. Besides, individual or organizational computer users need to apply some powerful firewall software to keep virus and malicious cyber intrusion away from the computer in use. On some occasions, wireless network is necessary to be used to secure the transmission of information against unauthorized intrusion. Among the mentioned precautions the most important one, however, might be to stay aware of various dangers from the web.

Passage 3: Computer ethics in the digital age

In this era computers exist everywhere, and technically speaking, they seem omnipotent. In human society, however, the application of any technology is not boundless. Then what do you suppose controls the way computers can be used? If you think it must be laws that control the operation and application of computer technology, you are definitely wrong! Obviously, the rapid development of this technology makes it very difficult for our legal system to keep up. Therefore, what we can rely on today as the essential element to control the way computers are used is ethics (See Figure 9. 4).

Figure 9. 4 Computer ethics are to be followed

Ethics is concerned with values, morals and standards by which human conduct can be judged as good or bad, right or wrong. in modern society computer ethics provide guidelines for how to use computers in a morally acceptable way. While using computers we should abide by the ethics meanwhile we are entitled to ethical treatment. As individual computer users, we enjoy the right to keep our personal privacy, information of credit ratings and medical histories, for instance, from falling into unauthorized parties. On the other side, we should restrict our conduct within the guidelines for computer ethics. These guidelines include but not be limited to the following (See Figure 9. 5).

1. Copyright and digital rights protection

Copyright is a legal term that gives content producers the right to control use and distribution of their work. Paintings, books, music, films, brands, and even video games are common items that can be copyrighted and hence legally protected. Computer software piracy causes big loss to legal software industry through unauthorized copying and distribution of original software. In addition to moral ethics, digital or software copyright laws are also considered necessary and effective in cracking down copyright piracy. You'd better be aware that selling, copying, downloading copyright-protected materials including music and videos are all illegal. If you want these materials, buy, borrow or log on authorized sites to access them, otherwise you'll be punished for copyright violation action.

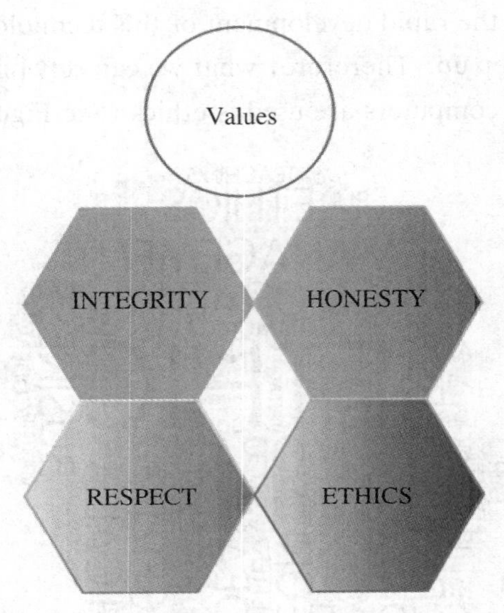

Figure 9. 5 Principles of computer ethics

2. Plagiarism is ethically and legally prohibited

The practice of representing someone else's words, ideas, or data without quotation, citation or acknowledgement is plagiarism. It's not only an act of intellectual dishonesty, but also an illegal conduct.

In digital era today, copying and pasting contents on web pages are much easier, yet you'll get into trouble when you use these contents directly in your books, papers or reports without citing the source. It is because identifying and catching plagiarists is becoming increasingly easy. By comparing a wide range of publicly available electronic materials, files and web contents, internet is incredibly effective in examining, detecting, and recognizing whether a paper is related to plagiarism.

9.2 Writing: Introduction to notice

Introduction to notice

There are several different definitions listed under the diction of notice. Under the situation of applied writing, notice means a formal announcement or notification of something, such as a meeting, activity, or a change in agenda.

Notice is one of the most commonly used form of communication. Usually, it is applied under the circumstance when the superior management personnel announce a decision, an arrangement, or call for a meeting. Notices are also issued within organizations of the same administrative status to discuss some affairs.

Ⅰ. Getting to know notice

1. General format

As a type of applied writing, notice is a simplified form of letter, it also has two general formats: block format and indented format. Notice has the features of being concise, brief and clear. Writing a notice does not require much lexical and rhetoric devices. Being to the point is the most important.

2. Layout of notice

In order to convey information appropriately to the notified, writing of notice should follow a certain layout. The text of notice can be organized in two general types. One is bulletin-like, informing people concerned, such as students, teachers, members, readers, audiences and so on. Another is letter-like, a more formal one to inform relevant recipients.

Generally speaking, the constituents of a notice of meeting include: title, event, time, place, participants, ways of participation, cautions, signature, notice issuer and date (See Figure 9.6).

Usually, the title of notice is put as "notice" or "announcement", and it is located at the top middle position. Sometimes the notifier's name is put right below the title, yet in some cases it can be put above the title, or put at the right corner at the end of the body.

The body of a notice is supposed to include event, time, place, expected participants, ways to participate, as well as dos and don'ts. In bulletin-like notice, parts of salutation and complimentary close are not needed, while in letter-like notice there is usually a line to indicate name(s) of the notified party at the left-hand position below the title.

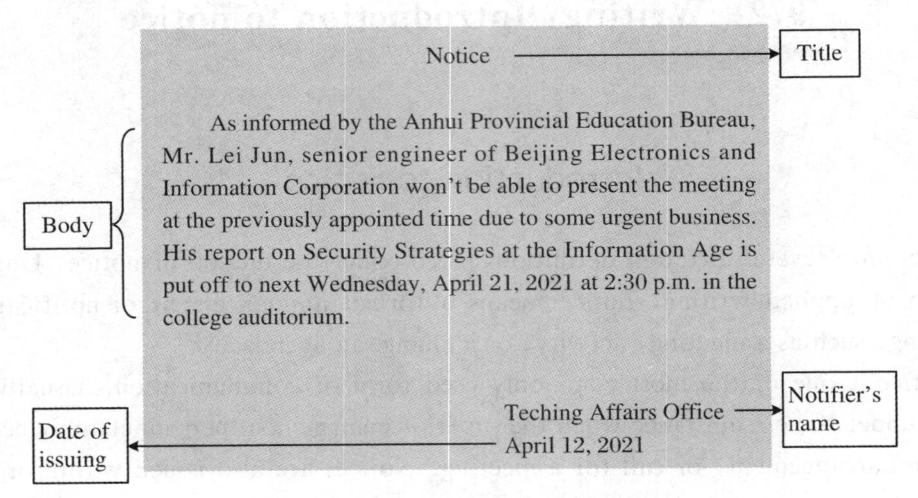

Figure 9.6　Constituents of a notice

Time for the issuing of notice is generally located at the right corner, a space below the notifier's name.

The individual or organization that issue the notice and the notified is described in the third person, but the second person should be applied if a salutation is used at the beginning of the notice. To better showcase the layout of a notice, the following is an example for a bulletin-like notice (a notice to put off a meeting).

Ⅱ. Sample Notice: A Notice of Meeting

Notice

We are going to have a company-wide meeting on how to ensure computer and internet safety. We invited a technician from Xunfei Information Institute who will report the importance of privacy and security for computer and internet users, as well as how to protect the cyber security of a business. Afterwards, discussion on this topic is to be arranged. The meeting will be held at 2:30 p.m. on Wednesday, September 15 in the conference room.

All employees are required to attend the meeting on time. Please be prepared to take notes and bring your questions.

Administrative Office

September 8, 2022

To summarize, you should include the following information in writing a notice, and pay attention to using the appropriate layout:

- State the background information of the notice
- Clarify the specific activity, arrangement of event
- State the attendee of the activity or event, or recipients of the notice
- Specify the date, time and place of the activity or event
- Clarify some details which demands special attention

9.3 Careers in IT

IT security specialist

IT security specialists' primary duties lie in: working in various industries to take and maintain digital protective measures on both intellectual and physical digital property as well as data that an organization possesses. They help companies make feasible and effective plans in case information gets hacked from their networks and servers. They are professionals who also create strategies to troubleshoot computer and system related problems.

Requirements for the position: A bachelor's degree or professional certification is often required. Academic background may involve courses such as math, programming and operating systems. Certifications offered by the Information Systems Security Certification Consortium (ISC2) is preferential.

Career prospect: Computer security specialist careers have a huge demand in the job market. These technical experts often come with large compensation packages. According to reports of the Bureau of Labor Statistics (BLS), the median salary level is far exceed average standard of all careers. Security specialists enjoy strong job prospects. Based on the BLS prediction that from 2020—2030, there will be a 33% job growth in this career.

Chief Information Security Officer (CISO)

The Chief Information Security Officer is normally an executive position of middle level whose responsibility is to manage a company's or organization's IT security division, and to maintain the security of personnel, information and physical assets. CISOs are usually responsible for their employers' demands on planning, coordinating and directing all computer, network and data security. At the same time, CISOs work directly with the management to identify and guarantee an organizations' and customers' cybersecurity requests. In addition, the CISOs are usually saddled with the responsibility of forming an effective staff team of security professionals.

Requirements for the position: The CISOs' responsibilities means that the position requires the applicant for this position have a strong background in IT security architecture and strategy. Meanwhile, the prospective CISOs are also expected to be equipped with effective communication, leadership and human resource skills.

Career prospects: This career enjoys a booming prospect for the fast development of IT industry. Being a high position in management, the career of CISOs is among the most high-paying ones. Besides lucrative paycheck the career of CISOs also provides smooth professional advancement suppose the job-holder can promote his/her educational background to a more satisfying level.

Cybersecurity Engineer

Cybersecurity engineers help businesses by protecting their computer and networking systems from potential hackers and cyber-attacks. They safeguard sensitive data of a business from hackers and cyber-criminals who often create new ways to infiltrate sensitive databases. Roles and responsibilities of a cybersecurity engineer include:

- Planning, implementing, managing, monitoring, and upgrading security measures for the protection of the organization's data, systems, and networks.
- Troubleshooting security and network problems.
- Responding to all system and/or network security breaches.
- Ensuring that the organization's data and infrastructure are protected by enabling the appropriate security controls.

- Participating in the change management process.
- Testing and identifying network and system vulnerabilities.
- Daily administrative tasks, reporting, and communication with the relevant departments in the organization.

9.4 Words and phrases

abide
anti-spyware
attack
browser
browser cache
computer security
cookies
copyright
cyber
cyber bullying
cybercrime
DoS (Denial of Service)
ethics
exaggeration
hack
malware
online identity
permission
phishing
plagiarism
positive
privacy
prohibit
protect
rogue Wi-Fi hotspots
scam
sensitive personal information
spoofing
spyware
unauthorized
Virtual Private Networks (VPNs)

9.5 Exercises

Ⅰ. Matching

Match each numbered item with the most closely related lettered item. Write your answers in the spaces provided.

a. privacy ________(1) wide range of programs that secretly record and report an individual's activities on the Internet

b. malware ________(2) malicious programs that damage or disrupt a computer system

c. phishing ________(3) infected computers that can be remotely controlled

d. plagiarism ________(4) a will-less and speechless human (as in voodoo belief and in fictional stories) held to have died and been supernaturally reanimated. How to use zombie in a sentence

e. spyware ________(5) the quality or state of being apart from company or observation: seclusion, freedom from unauthorized intrusion

f. security ________(6) an ethical issue relating to using another person's work and ideas as your own without giving credit to the original source

g. cyberbullying ________(7) a cybercrime in which a target or targets are contacted by email, telephone or text message by someone posing as a legitimate institution to lure individuals into providing sensitive data such as personally identifiable information, banking and credit card details, and passwords

h. zombies ________(8) an ethical issue relating to using another person's work and ideas as your own without giving credit to the original source

i. Trojan Horse ________(9) when bullying occurs in the online world — through computers, cell phones or other electronic devices — it's referred to it as bullying online or cyberbullying

Ⅱ. Written practice

(1) Describe what is computer privacy and security, as well as how to protect your personal data and the safety of your computer.

(2) Explain what is computer ethics and what are the principles of computer ethics.

(3) Write an English notice according to the given situation and information.

The Teaching Affairs Office of a college is going to call for a college-wide meeting in an attempt to help students guard against violation of intelligence property and plagiarism during the process of dissertation paper writing. The attendees include deans of all schools and department, all faculty, as well as the undergraduates and graduates in senior year.

Date: Oct. 19, 2021.

Time: 5 p.m.

Place: the lecture room on the second floor of the college library.

Ⅲ. Open-ended

On a separate sheet of paper, respond to each question or statement.

(1) Define privacy, and discuss the impact of large databases, private networks, the Internet, and the web.

(2) Define and discuss online identity and the major privacy laws.

(3) Define security. Define computer crime and the impact of malicious programs, including viruses, worms, Trojan horses, and zombies, as well as denial of service attacks, rogue Wi-Fi hotspots, data manipulation, identity theft, Internet scams, and cyberbullying.

(4) Discuss ways to protect computer security including restricting access, encrypting data, anticipating disasters, and preventing data loss.

(5) Define ethics, and describe copyright law and plagiarism.

Chapter 10 Errors, debug and test

Learning objectives

After you have read this chapter, you should be able to:

- ☆ Explain the errors and the bugs in programming.
- ☆ Explain how to debug and test program.
- ☆ Compare different errors, including syntax error, runtime error, logic error, and the others.
- ☆ Describe the differences between debugging and testing in programming.
- ☆ Describe careers in IT, including quality assurance tester and programmer.
- ☆ Identify software testing, including test case, test data and test plan.

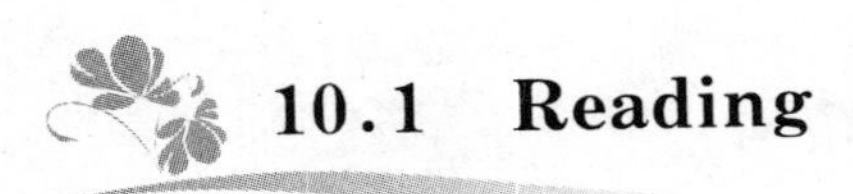

10.1 Reading

Passage 1: Errors in programming

Anyone involved in computer programming, even (perhaps especially) beginners are going to encounter errors and bugs of various types that force them to hunt down that culprit bit of code and make the necessary adjustments (See Figure 10.1). No matter how smart or how careful you are, errors are your constant companion. With practice, you will get slightly better at not making errors, and much, much better at finding and correcting them.

To understand debugging and testing more intuitively, lets first consider learning about different types of error that occurs while programming.

Figure 10.1 Computer programming

Ⅰ. What is a bug?

The term bug refers to any shortcoming in a software system that causes it to behave in unexpected and undesirable ways. These could range from irrational or incorrect responses, unpredictable failures, system crashes, etc. Essentially, it is a programming error leading to software malfunction, which has been detected before the website or app is deployed to production.

Bugs can be of multiple types. A few of them would be:

- Bugs affecting algorithms.
- Bugs affecting logic, e.g., infinite loops.
- Bugs emerging due to uninitialized variables.

Ⅱ. What is an error?

The term error refers to coding or programming mistake that usually shows up due to incorrect syntax or faulty loops. Errors emerge from the source code itself, caused by inconsistencies or outright fallacies in the internal code structure. They are anomalies triggered by misconceptions, oversights, or misunderstandings from the developer's (engineers, testers, analysts, etc.) end. The differences between bug and error are described as following (See Table 10.1):

Table 10.1 Difference between bug and error

Bug	Error
occurs due to shortcomings in the software system	occurs due to some mistake or misconception in the source code
detected before the software is pushed to production	detected when code is to be compiled and fails to do so
may be caused by human oversight or non-human causes like integration issues, environmental configuration, etc.	is caused by human oversight

Ⅲ. The common errors

There are three kinds of errors: syntax errors, runtime errors, and logic errors.

1. Syntax errors

Computer languages have their own specialized grammar rules. When these rules aren't followed (for example, the developer omits the parentheses while writing code), a syntax error prevents the application from running. In other words, the computer doesn't know what you want it to do. At this time, a syntax error is raised.

For example, you may have incorrect punctuation, or may be trying to use a variable that hasn't been declared and an extra/missing bracket at end of a function. Syntax errors are the easiest to find and correct. The compiler will tell you where it got into trouble, and its best guess as to what you did wrong. Usually, the error is on the exact line indicated by the compiler, or the line just before it. However, if the problem is incorrectly nested braces, the actual error may be at the beginning of the nested block. A good Integrated Development Environment (IDE) usually points out any syntax errors to the programmer (See Figure 10.2).

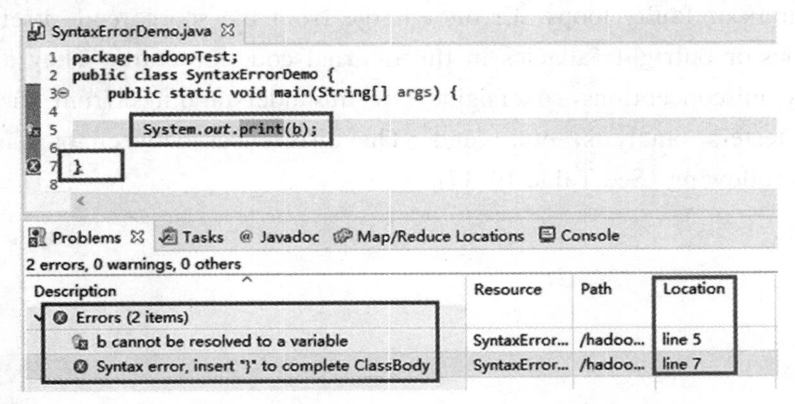

Figure 10.2 Example syntax errors in java

2. Runtime errors

These bugs occur when the code "won't play nice" with another computer, even if it worked perfectly fine on the developer's own computer. These errors are especially frustrating because they directly impact the end user and make the application appear unreliable or even completely broken.

For example, if there are no syntax errors, Java may detect an error while your program is running. You will get an error message telling you the kind of error, and a stack trace that tells not only where the error occurred, but also what other method or methods you were in.

```
Exception in thread "main" java.lang.NullPointerException
        atCar.placeInCity (Car.java:25)
        at City.< init>  (City.java:38)
        atCity.main (City.java:49)
```

It says that a NullPointerException was detected in the method "placeInCity" at line 25 in Car. java, which was called from the constructor for City at line 38 in City. java, which was called from the main method at line 49 in City. java. Sometimes there will be additional lines describing methods in the Java system itself; you can ignore these.

Runtime errors are intermediate in difficulty. Java tells you where it discovered that your program had gone wrong, but you need to trace back from there to figure out where the problem originated.

3. Logic errors

These errors can be extremely frustrating to deal with because nothing is inherently wrong with the code: the developer just didn't program the computer to do the correct thing. The system, of course, has no idea what your program is supposed to do, so it provides no additional information to help you find the errors. In fact, a logic error caused by miscalculations between American and English units caused NASA to lose a spacecraft in 1999.

Ways to track down a logic error include:

• Think about what the program must have done in order to produce the results it did. This will lead you to where the error must have occurred.

• Put in print statements to help you figure out what the program is actually doing.

• Use a debugger to step through your program and watch what it does.

Consider the example below (See Figure 10. 3), can you find errors?

The above lists are 3 basic types of errors. In addition, there are interface errors, resource errors and arithmetic errors, etc. (See Table 10. 2).

```
1  package hadoopTest;
2
3  public class LogincErrorDemo {
4      public static void main(String[] args) {
5          int Hours = 10;
6          double Hourly_Rate = 5.50;
7          double Pay = 10 * 6.50;
8          System.out.print(Pay);
9      }
10 }
```

Figure 10. 3 Example for login error

Table 10. 2 The common errors

Errors	Descriptions
Interface errors	These bugs typically happen when the inputs the software receives do not conform to the accepted standards. When handled incorrectly, these errors can look like errors on your side even when they're on the caller's side, and vice versa.
Resource errors	Sometimes, a program can force the computer it's running on to attempt to allocate more resources (processor power, random access memory, disk space, etc.) than it has. This results in the program becoming bugged or even causes the entire system to crash.
Arithmetic errors	These errors are just like logic errors, but with mathematics. For example, a division equation may require the computer to divide by zero. Since this is mathematically impossible, it results in an error that prevents the software from working correctly.

Passage 2: Debugging and testing

Ⅰ. Debugging VS Testing

Debugging is the process of finding errors and removing them from a computer program, otherwise they will lead to failure of the program. Even after taking full care during program design and coding, some errors may remain in the program and these errors appear during compilation or linking or execution. Debugging is generally done by program developer.

Testing is performed to verify that whether the completed software package functions or works according to the expectations defined by the requirements. Testing is generally performed by testing team which repetitively executes program with intent to find error. After testing, list of errors and related information is sent to program developer or development team.

Major differences between debugging and testing are pointed below in Table 10.3.

Table 10.3 Major differences between debugging and testing

Compare items	Debugging	Testing
What	It is the process of fixing errors.	It is the process of finding as many errors as possible.
When	Debugging is done during program development phase.	Testing is done during testing phase which comes after development phase.
Who	Debugging is done by program developer.	Testing is generally carried out by separate testing team rather than program developer.

Ⅱ. Best practices

1. Dealing with errors — debugging

When an error or bug has been identified, it needs to be corrected. Errors can often be fixed when program code is being written, and IDEs usually provide debugging tools which help identify these errors. Typical tools are:

- Single-step — programmers can step through the program one instruction at a time. The values of variables can be seen as they are processed and errors identified.
- Set break points — when the program is halted and the programmer can investigate the values of variables in the previous instructions. This is useful when the programmer suspects that the error is in a section of the program.
- Watchers — enable the programmer to watch for things like variables and program flow.
- Be sure to be familiar with the debugging tools in the programming language used.

2. Evaluating programs

Once a program has been written, the programmer may want to evaluate the strengths and weaknesses of a program. There are a number of criteria that can be used to decide how successful a program is:

- Is the program easy to use?
- Does the program meet all of the original requirements?
- Is the code maintainable?
- Is the program efficient?

The answers will help the programmer evaluate what could be done better next time or to refine the program, following evaluation.

3. Make trace tables

One way to test short programs is to carry out a dry run using paper. A dry run involves creating a trace table that includes all the variables a program contains. Whenever the value of a variable changes, the change is indicated in the trace table.

> **Trace tables** help programmers to determine the point in a program where a logic error has occurred.

Consider this simple pseudo-code program.

```
1  SET total TO 0
2  FOR count FROM 1 TO 3 DO 3
3  RECEIVE number FROM (INTEGER) KEYBOARD
4  SET total TO total +  number
5  END FOR
```

Each instruction has been given a line number -1 to 5. The program has three variables — total, count and number — which are put into a trace table.

Next, the program is tested using test data. If the numbers 5, 7 and 9 are input, the resulting total should be 21.

The instruction number is added to the trace table. If a variable changes with that instruction, the new variable value is written in the appropriate box (See Table 10.4).

Table 10.4 Sample of trace table

Instruction	Total	Count	Number
1	0		
2		1	
3			5
4	5		
5			
2		2	
3			7
4	12		
5			
2		3	
3			9
4	21		
5			

At each step, the programmer is able to see if, and how, a variable is affected.

Trace tables are extremely useful because they enable programmers to compare what the value of each variable should be against what a program actually produces. Where the two differ is the point in the program where a logic error has occurred.

10.2 Writing: How to write test plan

When first written, many programs contain bugs. Syntax errors and runtime errors are usually quickly removed, but it can take a long time to deduce where a logic error lies and why. The purpose of testing is to help programmers remove such bugs and to ensure that the program functions as intended.

Ⅰ. Test data

Test data is data that is used to test whether or not a program is functioning correctly. The test data is input, the program is processed and the output is confirmed.

Whenever possible, test data should cover a range of possible and impossible inputs, each designed to prove a program works or to highlight any flaws. Three types of data are:

- Normal data — sensible, valid data that the program should accept and be able to process.
- Boundary data — valid data that falls at the boundary of any possible ranges.
- Erroneous data — invalid data that the program cannot process and should not accept.

Ⅱ. Test plans

Testing requires a test plan. This is a list of all the tests to be used to ensure the program functions as intended. The list should include several examples of normal, boundary and erroneous data.

Tests are laid out in a test plan which might contain (See Figure 10.4):

- The test numbers.
- A description of what the test intends to check.
- The test data being used.
- The type of test-normal, boundary or erroneous.
- Expected outcome.
- Actual outcome.

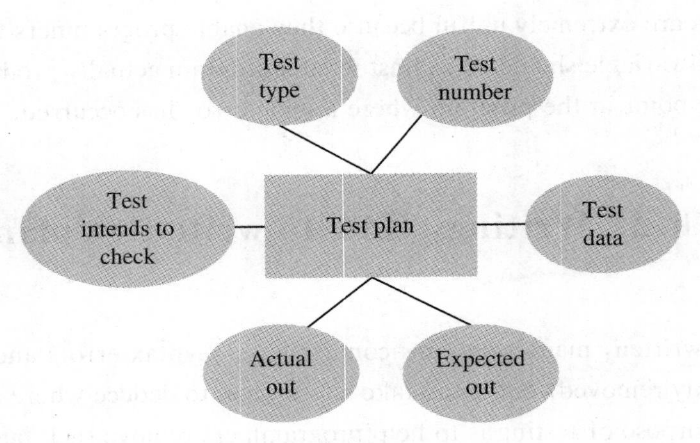

Figure 10.4 Test plan entity

Ⅲ. Sample: Make a test plan

Consider this simple pseudo-code program, which asks a user to input a number from 1 to 10:

```
SET valid TO False
  WHILE valid =  False DO
      SEND 'Enter a number from 1 to 10' TO DISPLAY
      RECEIVE number FROM (INTEGER) KEYBOARD
      IF number < 1 OR number > 10 THEN
        SEND 'Number outside the range 1 to 10.
        Enter another number' TO DISPLAY
      ELSE
      SET valid TO True
  END WHILE
  SEND 'Number entered is', number TO DISPLAY
```

This program could be tested using the following normal, boundary and erroneous data, see the below Table 10.5.

Test plans should be created before programming starts so that they reflect all the requirements of the program design.

Programmers should run as many tests as is sensible. Many large programs, especially games, contain bugs simply because it may not be possible to test every possible input or action.

Table 10.5 Test plan samples

Test number	Description	Test data	Test type	Expected	Actual
1	Test that a possible number is accepted	5	Normal	Data is accepted	Data is accepted
2	Test the lower boundary	1	Boundary	Data is accepted	Data is accepted
3	Test the upper boundary	10	Boundary	Data is accepted	Data is accepted
4	Test that the program does not accept a number less than 1	−5	Erroneous	Data is not accepted	Data is not accepted
5	Test that the program does not accept a number greater than 10	20	Erroneous	Data is not accepted	Data is not accepted

10.3 Careers in IT

Quality assurance tester

Primary duties: Quality assurance testers are technicians or engineers who check software products to see if they're up to industry standards and free of any issues. This role is common for gaming systems, mobile applications and other technology that needs further testing and maintenance when recommended.

Requirements: Many quality assurance testers have a Bachelor's Degree in software design, engineering or computer science. Testers can work on different software for IT companies, which may influence what degree or specialization they pursue. These professionals should also have excellent time management and communication skills to help document test cases.

Computer programmer

Primary duties: A computer programmer is someone who writes new computer software using coding languages like HTML, JavaScript and CSS. Video game software can be updated to improve online gameplay, which is an opportunity for programmers to troubleshoot problems experienced by gamer after the game is released to the general public.

Requirements: A programmer typically completes a Bachelor's Degree in computer science and an internship to build their skills. Certifications are also strongly encouraged, and there are many coding academies to choose from.

10.4 Words and phrases

boundary data
computer programming
criteria
culprit
debug
debugger
dry run
erroneous data
errors and bugs
evaluate
exception
failure
grammar rules
hunt down
incorrect punctuation
incorrect response
IDE
intuitively
invalid
irrational
logic errors
malfunction
normal data
NullPointerException
occur
pseudo-code
requirement
runtime errors
set break points
single-step
stack trace
syntax errors
system crash
test data
test plan
testing team
trace table
undesirable
unexpected
unpredictable
valid
variable

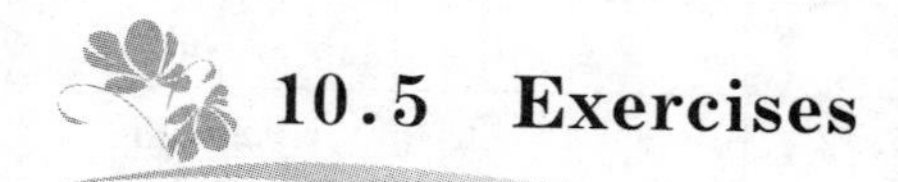

10.5 Exercises

Ⅰ. Matching

Match each numbered item with the most closely related lettered item. Write your answers in the spaces provided.

a. criteria ________(1) an error in a program

b. syntax errors ________(2) a manual test of program code carried out by the programmer, where they manually calculate values of variables as they work through code line by line

c. test data ________(3) data input when testing to see if the program produces the expected results

d. test plan ________(4) integrated development environment — a piece of software used to write computer programs

e. instruction ________(5) a list of what is to be tested and how it is to be tested

f. IDE ________(6) the process of removing errors from a program

g. normal data ________(7) a set of rules or conditions that must be met

h. dry run ________(8) a single action that can be performed by a computer processor

i. debug ________(9) error in a program resulting from code not following syntax rules governing how to write statements in a programming language

j. bug ________(10) sensible, possible data that the program should accept and be able to process

Ⅱ. Written practice

Given three sides, check whether triangle is valid or not. A triangle is valid if sum of its two sides is greater than the third side. If three sides are a, b and c, then three conditions should be met.

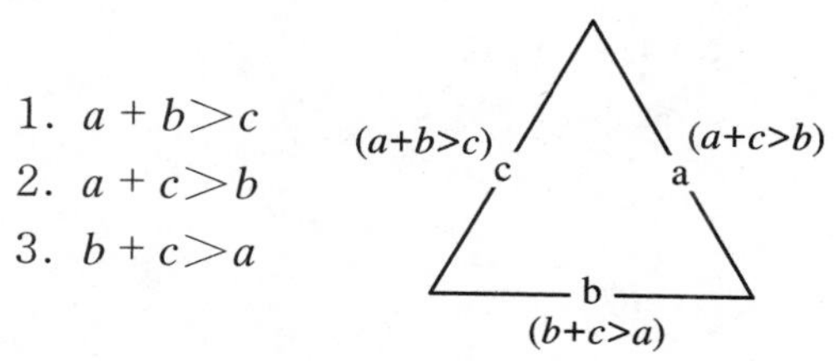

Test plan samples

Test number	Description	Side *a*	Side *b*	Side *c*	Expected outcome	Actual outcome
1	normal data	3	4	5	valid	
2	erroneous data	1	2	3	invalid	

Please refer to "How to write test plan" sections to design the test plan table of "A triangle is valid or not" as above sample table.

Ⅲ. Open-ended

On a separate sheet of paper, respond to each question or statement.

(1) Describe the error and bug.
(2) Describe how to debugging the logic errors.
(3) What's the difference between debugging and testing?

Appendix Ⅰ Computer glossary

Word	Description
application	Computer software that performs a task or set of tasks, such as word processing or drawing. Applications are also referred to as programs.
ASCII	American Standard Code for Information Interchange, an encoding system for converting keyboard characters and instructions into the binary number code that the computer understands.
bandwidth	The capacity of a networked connection. Bandwidth determines how much data can be sent along the networked wires. Bandwidth is particularly important for Internet connections, since greater bandwidth also means faster downloads.
binary code	The most basic language a computer understands, it is composed of a series of 0s and 1s. The computer interprets the code to form numbers, letters, punctuation marks, and symbols.
bit	The smallest piece of computer information, either the number 0 or 1. In short they are called binary digits.
boot	To start up a computer. Cold boot means restarting computer after the power is turned off. Warm boot means restarting computer without turning off the power.
browser	Software used to navigate the Internet. Google Chrome, Firefox, Netscape Navigator and Microsoft Internet Explorer are today's most popular browsers for accessing the World Wide Web.
bug	A malfunction due to an error in the program or a defect in the equipment.
byte	Most computers use combinations of eight bits, called bytes, to represent one character of data or instructions. For example, the word cat has three characters, and it would be represented by three bytes.
cache	A small data-memory storage area that a computer can use to instantly re-access data instead of re-reading the data from the original source, such as a hard drive. Browsers use a cache to store web pages so that the user may view them again without reconnecting to the Web.
CAD/CAM	Computer Aided Drawing/Computer Aided Manufacturing. The instructions stored in a computer that will be translated to very precise operating instructions to a robot, such as for assembling cars or laser-cutting signage.

Continued

Word	Description
CD-ROM	Compact Disc Read-Only Memory, an optically read disc designed to hold information such as music, reference materials, or computer software. A single CD-ROM can hold around 640 megabytes of data, enough for several encyclopaedias. Most software programs are now delivered on CD-ROMs.
chat	Typing text into a message box on a screen to engage in dialogue with one or more people via the Internet or other network.
chip	A tiny wafer of silicon containing miniature electric circuits that can store millions of bits of information.
client	A single user of a network application that is operated from a server. A client/server architecture allows many people to use the same data simultaneously. The program's main component (the data) resides on a centralized server, with smaller components (user interface) on each client.
cookie	A text file sent by a Web server that is stored on the hard drive of a computer and relays back to the Web server things about the user, his or her computer, and/or his or her computer activities.
CPU	Central Processing Unit. The brain of the computer.
cracker	A person who breaks in to a computer through a network, without authorization and with mischievous or destructive intent.
crash	A hardware or software problem that causes information to be lost or the computer to malfunction. Sometimes a crash can cause permanent damage to a computer.
cursor	A moving position-indicator displayed on a computer monitor that shows a computer operator where the next action or operation will take place.
cyberspace	Slang for internet. An international conglomeration of interconnected computer networks. Begun in the late 1960s, it was developed in the 1970s to allow government and university researchers to share information. The Internet is not controlled by any single group or organization. Its original focus was research and communications, but it continues to expand, offering a wide array of resources for business and home users.
Database	A collection of similar information stored in a file, such as a database of addresses. This information may be created and stored in a Database Management System (DBMS).
debug	Slang. To find and correct equipment defects or program malfunctions.
default	The predefined configuration of a system or an application. In most programs, the defaults can be changed to reflect personal preferences.

Continued

Word	Description
desktop	The main directory of the user interface. Desktops usually contain icons that represent links to the hard drive, a network (if there is one), and a trash or recycling can for files to be deleted. It can also display icons of frequently used applications, as requested by the user.
desktop publishing	The production of publication-quality documents using a personal computer in combination with text, graphics, and page layout programs.
directory	A repository where all files are kept on computer.
disk	Two distinct types. The names refer to the media inside the container: A hard disc stores vast amounts of data. It is usually inside the computer but can be a separate peripheral on the outside. Hard discs are made up of several rigid coated metal discs. Currently, hard discs can store 15 to 30 Gb (gigabytes). A floppy disc, 3.5 "square, usually inserted into the computer and can store about 1.4 megabytes of data. The 3.5" square floppies have a very thin, flexible disc inside. There is also an intermediate-sized floppy disc, trademarked Zip discs, which can store 250 megabytes of data.
disk drive	The equipment that operates a hard or floppy disc.
domain	Represents an IP (Internet Protocol) address or set of IP addresses that comprise a domain. The domain name appears in URLs to identify web pages or in email addresses.
domain name	The name of a network or computer linked to the Internet. Domains are defined by a common IP address or set of similar IP (Internet Protocol) addresses.
download	The process of transferring information from a web site (or other remote location on a network) to the computer. It is possible to download a file which include text, image, audio, video and many others.
DOS	Disk Operating System. An operating system designed for early IBM-compatible PCs.
drop-down menu	A menu window that opens vertically on-screen to display context-related options. Also called pop-up menu or pull-down menu.
DSL	Digital Subscriber Line, a method of connecting to the Internet via a phone line. A DSL connection uses copper telephone lines but is able to relay data at much higher speeds than modems and does not interfere with telephone use.
DVD	Digital Video Disc. Similar to a CD-ROM, it stores and plays both audio and video.
E-book	An electronic (usually hand-held) reading device that allows a person to view digitally stored reading materials.

Continued

Word	Description
Email	Electronic mail; messages, including memos or letters, sent electronically between networked computers that may be across the office or around the world.
emoticon	A text-based expression of emotion created from ASCII characters that mimics a facial expression when viewed with your head tilted to the left. Here are some examples: smiling, frowning, winking, crying.
encryption	The process of transmitting scrambled data so that only authorized recipients can unscramble it. For instance, encryption is used to scramble credit card information when purchases are made over the Internet.
Ethernet	A type of network.
Ethernet card	A board inside a computer to which a network cable can be attached.
file	A set of data that is stored in the computer.
firewall	A set of security programs that protect a computer from outside interference or access via the Internet.
folder	A structure for containing electronic files. In some operating systems, it is called a directory.
fonts	Sets of typefaces (or characters) that come in different styles and sizes.
freeware	Software created by people who are willing to give it away for the satisfaction of sharing or knowing they helped to simplify other people's lives. It may be free-standing software, or it may add functionality to existing software.
FTP	File Transfer Protocol, a format and set of rules for transferring files from a host to a remote computer.
Gigabyte (GB)	1024 megabytes. Also called gig.
glitch	The cause of an unexpected malfunction.
gopher	An Internet search tool that allows users to access textual information through a series of menus, or if using FTP, through downloads.
GUI	Graphical User Interface, a system that simplifies selecting computer commands by enabling the user to point to symbols or illustrations (called icons) on the computer screen with a mouse.
groupware	Software that allows networked individuals to form groups and collaborate on documents, programs, or databases.
hacker	A person with technical expertise who experiments with computer systems to determine how to develop additional features. Hackers are occasionally requested by system administrators to try and break into systems via a network to test security. The term hacker is sometimes incorrectly used interchangeably with cracker. A hacker is called a white hat and a cracker a black hat.

Continued

Word	Description
hard copy	A paper printout of what you have prepared on the computer.
hard drive	Another name for the hard disc that stores information in a computer.
hardware	The physical and mechanical components of a computer system, such as the electronic circuitry, chips, monitor, disks, disk drives, keyboard, modem, and printer.
home page	The main page of a Web site used to greet visitors, provide information about the site, or to direct the viewer to other pages on the site.
HTML	Hypertext Markup Language, a standard of text markup conventions used for documents on the World Wide Web. Browsers interpret the codes to give the text structure and formatting (such as bold, blue, or italic).
HTTP	Hypertext Transfer Protocol, a common system used to request and send HTML documents on the World Wide Web. It is the first portion of all URL addresses on the World Wide Web.
HTTPS	Hypertext Transfer Protocol Secure, often used in intracompany internet sites. Passwords are required to gain access.
hyperlink	Text or an image that is connected by hypertext coding to a different location. By selecting the text or image with a mouse, the computer jumps to (or displays) the linked text.
hypermedia	Integrates audio, graphics, and/or video through links embedded in the main program.
hypertext	A system for organizing text through links, as opposed to a menu-driven hierarchy such as Gopher. Most Web pages include hypertext links to other pages at that site, or to other sites on the World Wide Web.
icons	Symbols or illustrations appearing on the computer screen that indicate program files or other computer functions.
input	Data that goes into a computer device.
input device	A device, such as a keyboard, stylus and tablet, mouse, puck, or microphone, that allows input of information (letters, numbers, sound, video) to a computer.
instant messaging (IM)	A chat application that allows two or more people to communicate over the Internet via real-time keyed-in messages.
interface	The interconnections that allow a device, a program, or a person to interact. Hardware interfaces are the cables that connect the device to its power source and to other devices. Software interfaces allow the program to communicate with other programs (such as the operating system), and user interfaces allow the user to communicate with the program (e.g., via mouse, menu commands, icons, voice commands, etc.).

Continued

Word	Description
Internet	An international conglomeration of interconnected computer networks. Begun in the late 1960s, it was developed in the 1970s to allow government and university researchers to share information. The Internet is not controlled by any single group or organization. Its original focus was research and communications, but it continues to expand, offering a wide array of resources for business and home users.
IP (Internet Protocol) address	An Internet Protocol address is a unique set of numbers used to locate another computer on a network. The format of an IP address is a 32-bit string of four numbers separated by periods. Each number can be from 0 to 255 (i.e., 1.154.10.255). Within a closed network IP addresses may be assigned at random, however, IP addresses of web servers must be registered to avoid duplicates.
Java	An object-oriented programming language designed specifically for programs (particularly multimedia) to be used over the Internet. Java allows programmers to create small programs or applications (applets) to enhance Web sites.
JavaScript/ECMA script	A programming language used almost exclusively to manipulate content on a web page. Common JavaScript functions include validating forms on a web page, creating dynamic page navigation menus, and image rollovers.
Kilobyte (K or KB)	Equal to 1,024 bytes.
Linux	A UNIX-like, open-source operating system developed primarily by Linus Torvalds. Linux is free and runs on many platforms, including both PCs and servers. Linux is an open-source operating system, meaning that the source code of the operating system is freely available to the public. Programmers may redistribute and modify the code, as long as they don't collect royalties on their work or deny access to their code. Since development is not restricted to a single corporation, more programmers can debug and improve the source code faster.
laptop and notebook	Small, lightweight, portable battery-powered computers that can fit onto your lap. They each have a thin, flat, liquid crystal display screen.
Macro	A script that operates a series of commands to perform a function. It is set up to automate repetitive tasks.
Mac OS	An operating system with a graphical user interface, developed by Apple for Macintosh computers. Current System X. 1. (10) combines the traditional Mac interface with a strong underlying UNIX. Operating system for increased performance and stability.
megabyte (MB)	Equal to 1, 048, 576 bytes, usually rounded off to one million bytes (also called a meg).

Continued

Word	Description
memory	Temporary storage for information, including applications and documents. The information must be stored to a permanent device, such as a hard disc or CD-ROM before the power is turned off, or the information will be lost. Computer memory is measured in terms of the amount of information it can store, commonly in megabytes or gigabytes.
menu	A context-related list of options that users can choose from.
menu bar	The horizontal strip across the top of an application's window. Each word on the strip has a context sensitive drop-down menu containing features and actions that are available for the application in use.
merge	To combine two or more files into a single file.
Microprocessor	A complete central processing unit (CPU) contained on a single silicon chip.
minimize	A term used in a GUI operating system that uses windows. It refers to reducing a window to an icon, or a label at the bottom of the screen, allowing another window to be viewed.
modem	A device that connects two computers together over a telephone or cable line by converting the computer's data into an audio signal. Modem is a contraction for the process it performs: modulate/demodulate.
monitor	A video display terminal.
mouse	A small hand-held device, similar to a trackball, used to control the position of the cursor on the video display; movements of the mouse on a desktop correspond to movements of the cursor on the screen.
MP3	Compact audio and video file format. The small size of the files makes them easy to download and e-mail. Format used in portable playback devices.
multimedia	Software programs that combine text and graphics with sound, video, and animation. A multimedia PC contains the hardware to support these capabilities.
MS-DOS	An early operating system developed by Microsoft Corporation (Microsoft Disc Operating System).
network	A system of interconnected computers.
open source	Computer programs whose original source code was revealed to the general public so that it could be developed openly. Software licensed as open source can be freely changed or adapted to new uses, meaning that the source code of the operating system is freely available to the public. Programmers may redistribute and modify the code, as long as they don't collect royalties on their work or deny access to their code. Since development is not restricted to a single corporation more programmers can debug and improve the source code faster.

Continued

Word	Description
Operating System	A set of instructions that tell a computer on how to operate when it is turned on. It sets up a filing system to store files and tells the computer how to display information on a video display. Most PC operating systems are DOS (Disc Operated System) systems, meaning the instructions are stored on a disc (as opposed to being originally stored in the microprocessors of the computer). Other well-known operating systems include UNIX, Linux, and Windows.
output	Data that come out of a computer device. For example, information displayed on the monitor, sound from the speakers, and information printed to paper.
palm	A hand-held computer.
PC board	Printed Circuit board, a board printed or etched with a circuit and processors. Power supplies, information storage devices, or changers are attached.
PDA	Personal Digital Assistant, a hand-held computer that can store daily appointments, phone numbers, addresses, and other important information. Most PDAs link to a desktop or laptop computer to download or upload information.
PDF	Portable Document Format, a format presented by Adobe Acrobat that allows documents to be shared over a variety of operating systems. Documents can contain words and pictures and be formatted to have electronic links to other parts of the document or to places on the web.
Pentium chip	Intel's fifth generation of sophisticated high-speed microprocessors. Pentium means the fifth element.
peripheral	Any external device attached to a computer to enhance operation. Examples include external hard drive, scanner, printer, speakers, keyboard, mouse, trackball, stylus and tablet, and joystick.
Personal Computer (PC)	A single-user computer containing a Central Processing Unit (CPU) and one or more memory circuits.
petabyte	A measure of memory or storage capacity and is approximately a thousand terabytes.
platform	The operating system, such as UNIX, Macintosh, Windows, on which a computer is based.
plug and play	Computer hardware or peripherals that come set up with necessary software so that when attached to a computer, they are recognized by the computer and are ready to use.
pop-up menu	A menu window that opens vertically or horizontally on-screen to display context-related options. Also called drop-down menu or pull-down menu.
power PC	A competitor of the Pentium chip. It is a new generation of powerful sophisticated microprocessors produced from an Apple-IBM-Motorola alliance.

Continued

Word	Description
printer	A mechanical device for printing a computer's output on paper. There are three major types of printer.
program	A precise series of instructions written in a computer language that tells the computer what to do and how to do it. Programs are also called software or applications.
programming language	A series of instructions written by a programmer according to a given set of rules or conventions (syntax). High-level programming languages are independent of the device on which the application (or program) will eventually run; low-level languages are specific to each program or platform. Programming language instructions are converted into programs in language specific to a particular machine or operating system (machine language). So that the computer can interpret and carry out the instructions. Some common programming languages are: C, C++, Java, Python, and PHP.
puck	An input device, like a mouse. It has a magnifying glass with crosshairs on the front of it that allows the operator to position it precisely when tracing a drawing for use with CAD/CAM software.
pull-down menu	A menu window that opens vertically on-screen to display context-related options. Also called drop-down menu or pop-up menu.
push technology	Internet tool that delivers specific information directly to a user's desktop, eliminating the need to surf for it. PointCast, which delivers news in user-defined categories, is a popular example of this technology.
QuickTime	Audio-visual software that allows movie-delivery via the Internet and email. QuickTime images are viewed on a monitor.
RAID	Redundant Array of Inexpensive Disks, a method of spreading information across several disks set up to act as a unit, using two different techniques: Disk striping — storing a bit of information across several discs (instead of storing it all on one disc and hoping that the disc doesn't crash); Disk mirroring — simultaneously storing a copy of information on another disc so that the information can be recovered if the main disc crashes.
RAM	Random Access Memory, one of two basic types of memory. Portions of programs are stored in RAM when the program is launched so that the program will run faster. Though a PC has a fixed amount of RAM, only portions of it will be accessed by the computer at any given time. Also called memory.
right-click	Using the right mouse button to open context-sensitive drop-down menus.
ROM	Read-Only Memory, one of two basic types of memory. ROM contains only permanent information put there by the manufacturer. Information in ROM cannot be altered, nor can the memory be dynamically allocated by the computer or its operator.

Continued

Word	Description
scanner	An electronic device that uses light-sensing equipment to scan paper images such as text, photos, and illustrations and translate the images into signals that the computer can then store, modify, or distribute.
search engine	Software that makes it possible to look for and retrieve material on the Internet, particularly the Web. Some popular search engines are Baidu, Google, etc.
server	A computer that shares its resources and information with other computers, called clients, on a network.
shareware	Software created by people who are willing to sell it at low cost or no cost for the gratification of sharing. It may be freestanding software, or it may add functionality to existing software.
software	Computer programs, also called applications.
spider	A process search engines use to investigate new pages on a web site and collect the information that needs to be put in their indices.
spreadsheet	Software that allows one to calculate numbers in a format that is similar to pages in a conventional ledger.
storage	Devices used to store massive amounts of information so that it can be readily retrieved. Devices include RAIDs, CD-ROMs, DVDs.
streaming	Taking packets of information (sound or visual) from the Internet and storing it in temporary files to allow it to play in continuous flow.
stylus and tablet	An input device similar to a mouse. The stylus is pen shaped. It is used to draw on a tablet (like drawing on paper) and the tablet transfers the information to the computer. The tablet responds to pressure. The firmer the pressure used to draw, the thicker the line appears.
surfing	Exploring the Internet.
surge protector	A controller to protect the computer and make up for variances in voltage.
Telnet	A way to communicate with a remote computer over a network.
Trackball	Input device that controls the position of the cursor on the screen; the unit is mounted near the keyboard, and movement is controlled by moving a ball.
Terabytes (TB)	A thousand gigabytes.
UNIX	A very powerful operating system used as the basis of many high-end computer applications.
upload	The process of transferring information from a computer to a web site (or other remote location on a network). To transfer information from a computer to a web site (or other remote location on a network).
URL	Uniform Resource Locator. The protocol for identifying a document on the Web. A Web address (e. g. , www. tutorialspoint. com). A URL is unique to each user.

Continued

Word	Description
UPS	Universal Power Supply or Uninterruptible Power Supply. An electrical power supply that includes a battery to provide enough power to a computer during an outage to back-up data and properly shut down.
USB	A multiple-socket USB connector that allows several USB-compatible devices to be connected to a computer.
user friendly	A program or device whose use is intuitive to people with a non-technical background.
video teleconferencing	A remote "face-to-face chat", when two or more people using a webcam and an Internet telephone connection chat online. The webcam enables both live voice and video.
Virtual Reality (VR)	A technology that allows one to experience and interact with images in a simulated three-dimensional environment. For example, you could design a room in a house on your computer and actually feel that you are walking around in it even though it was never built. (The Holodeck in the science-fiction TV series Star Trek : Voyager would be the ultimate virtual reality.) Current technology requires the user to wear a special helmet, viewing goggles, gloves, and other equipment that transmits and receives information from the computer.
Virus	An unauthorized piece of computer code attached to a computer program or portions of a computer system that secretly copies itself from one computer to another by shared discs and over telephone and cable lines. It can destroy information stored on the computer, and in extreme cases, can destroy operability. Computers can be protected from viruses if the operator utilizes good virus prevention software and keeps the virus definitions up to date. Most viruses are not programmed to spread themselves. They have to be sent to another computer by email, sharing, or applications. The worm is an exception, because it is programmed to replicate itself by sending copies to other computers listed in the email address book in the computer. There are many kinds of viruses, for example: (1) Boot viruses place some of their code in the start-up disk sector to automatically execute when booting. Therefore, when an infected machine boots, the virus loads and runs. (2) File viruses attached to program files (files with the extension. exe). When you run the infected program, the virus code executes. (3) Macro viruses copy their macros to templates and/or other application document files. (4) Trojan Horse is a malicious, security-breaking program that is disguised as something being such as a screen saver or game. (5) Worm launches an application that destroys information on your hard drive. It also sends a copy of the virus to everyone in the computer's email address book.

Continued

Word	Description
WAV	A sound format (pronounced wave) used to reproduce sounds on a computer.
Webcam	A video camera/computer setup that takes live images and sends them to a Web browser.
Window	A portion of a computer display used in a graphical interface that enables users to select commands by pointing to illustrations or symbols with a mouse. "Windows" is also the name Microsoft adopted for its popular operating system.
World Wide Web ("WWW" or "the Web")	A network of servers on the Internet that use hypertext-linked databases and files. It was developed in 1989 by Tim Berners-Lee, a British computer scientist, and is now the primary platform of the Internet. The feature that distinguishes the Web from other Internet applications is its ability to display graphics in addition to text.
Word processor	A computer system or program for setting, editing, revising, correcting, storing, and printing text.
WYSIWYG	What You See Is What You Get. When using most word processors, page layout programs, and web page design programs, words and images will be displayed on the monitor as they will look on the printed page or web page.

Appendix Ⅱ Computer abbreviations

Abbreviations	Chinese
3D (Three-Dimension)	
3GL (third-generation language)	
4GLs (fourth-generation languages)	
5G (5th Generation Mobile Communication Technology)	
5GL (fifth-generation language)	
ADSL (Asymmetric Digital Subscriber Line)	
AGI (Artificial General Intelligence)	
AI (Artificial Intelligence)	
AJAX (Asynchronous JavaScript and XML)	
ALU (Arithmetic and Logic Unit)	
ANI (Artificial Narrow Intelligence)	
ANN (Artificial Neural Networks)	
ANSI (American National Standards Committee)	
API (Application Program Interface)	
ASCII (American Standard Code for Information Interchange)	
ASI (Artificial Super Intelligence)	
ASP (Active Server Page)	
ATM (Asynchronous Transfer Mode)	
AVI (Audio Video Interleave)	
BIOS (Basic Input Output System)	
BMC (Baseboard Manager Controller)	
CAD (Computer-Aided Design)	
CAE (Computer-aided Engineering)	
CAM (Computer-aided Manufacturing)	
CAPP (Computer-aided Process Planning)	
CASE (Computer-aided Software Engineering)	

Continued

Abbreviations	Chinese
CCIE (Cisco Certified Internet work Expert)	
CCNA (Cisco Certified Network Associate)	
CCNP (Cisco Certified Network Professional)	
CCT (Cisco Certified Technician)	
CD (Compact Disk)	
CD ROM (Compact Disk ROM)	
CGI (Common Gateway Interface)	
CIM (Computer-integrated Manufacturing)	
CLI (Command-line Interface)	
CLR (Common Language Runtime)	
CLS (Common Language Specification)	
CPE (Customer Premises Equipment)	
CPU (Central Processing Unit)	
CRM (Customer Relationship Management)	
C-S (Client-server)	
CSS (Cascading Style Sheets)	
CV (Computer Vision)	
DBMS (Database Management System)	
DDoS (Distributed Denial of Service)	
DDR (Double Data Rate)	
DHTML (Dynamic Hypertext Markup Language)	
DIKW (Data Information Knowledge Wisdom)	
DLL (Dynamic Link Library)	
DMA (Direct Memory Access)	
DNS (Domain Name Server)	
DOM (Document Object Model)	
DoS (Denial of service)	
DRAM (Dynamic RAM)	
DVD (Digital Video Disk)	
DVI (Digital Video Interface)	

Continued

Abbreviations	Chinese
ECMA (European Computer Manufacturer's Association)	
EFI (Extensible Firmware Interface)	
EFS (Encrypting File System)	
ERP (Enterprise Resource Planning)	
FDDI (Fiber Distributed Data Interface)	
FTP File Transfer Protocol	
GB (Gigabyte)	
GIF (Graphics Interchange Format)	
GPL (General Public License)	
GPS (Global Positioning Satellite)	
GRUB (GRand Unified Bootloader)	
GUI (Graphic User Interface)	
HPC (High Performance Computing)	
HTML (HypertText Markup Language)	
HTTP (Hypertext Transfer Protocol)	
HTTPS (Hypertext Transfer Protocol over Secure Socket Layer)	
I/O (Input/Output)	
IaaS (Infrastructure as a Service)	
ICT (Information and Communications Technology)	
IDC (Internet Data Center)	
IDE (Integrated Development Environment)	
IE (Internet Explorer)	
IIS (Internet Information Server)	
IoT (Internet of Things)	
IP (Internet Protocol)	
IPC (Inter-Process Communication)	
ISO (International Standards Organization)	
ISP (Internet Service Provider)	
JPEG (Joint Photographic Experts Group)	
JPG (Joint Picture Group)	

Continued

Abbreviations	Chinese
KB (Kilobyte)	
KPIs (Key Performance Indicators)	
KVM (Kernel-based Virtual Machine)	
LAN (Local Area Network)	
MAC (Medium Access Control)	
MAN (Metropolitan Area Network)	
MBR (Master Boot Record)	
MIPS (Millions of Instructions executed Per Second)	
MIS (Management Information System)	
ML (Machine Learning)	
NAP (Network Access Protection)	
NFS (Network File System)	
NGI (Next Generation Internet)	
NIC (Network Interface Card)	
NLP (Natural Language Processing)	
NoSQL (Not only Structured Query Language)	
OA (Office Automatic)	
OOP (Object-Oriented Programming)	
OS (Operating System)	
OSI (Open System Interconnect Reference Model)	
P2P (Peer to Peer)	
PaaS (Platform as a Service)	
PAN (Personal Area Network)	
PB (Petabytes)	
PC (Personal Computer)	
PCI (Peripheral Component Interconnect)	
PDA (Personal Digital Assistant)	
PDF (Portable Document Format)	
PDL (Page Description Language)	
PHP (Personal Home Page)	

Continued

Abbreviations	Chinese
POP (Post Office Protocol)	
POSIX (Portable Operating System UNIX)	
POST (Power OnSelf Test)	
RAID (Redundant Arrays of Independent Disks)	
RAM (Random Access Memory)	
RDBMS (Relational Database Management System)	
RF (Radio Frequency)	
RISC (Reduced Instruction-Set Computer)	
ROM (Read Only Memory)	
RTF (Rich Text Format)	
SaaS (Software as a Service)	
SATA (Serial Advanced Technology Attachment)	
SCSI (Small Computer System Interface)	
SDK (Software Development Kit)	
SDLC (Software Development Life Cycle)	
SFTP (Secure File Transfer Protocol)	
SMTP (Simple Mail Transfer Protocol)	
SQL (Structured Query Language)	
SSH (Secure Shell)	
TB (Terabyte)	
TCP (Transmission Control Protocol)	
IP (Internet Protocol)	
TIFF (Tagged Image File Format)	
UDP (User Datagram Protocol)	
UEFI (Unified Extensible Firmware Interface)	
UNIX (Uniplexed Information and Computing System)	
UPS (Uninterruptible Power Supply)	
URL (Uniform Resource Locator)	
USB (Universal Serial Bus)	
UTP (Unshielded Twisted Pair)	

Continued

Abbreviations	Chinese
VGA (Video Graphics Array)	
VM (Virtual Machine)	
VOD (Video On Demand)	
VPN (Virtual Private Network)	
VRML (Virtual Reality Modeling Language)	
WAN (Wide Area Network)	
WAP (Wireless Application Protocol)	
Wi-Fi (Wireless Fidelity)	
WLAN (Wireless Local Area Network)	
WORM (Write Once, Read Many)	
WPA (Wi-Fi Protected Access)	
WWW (World Wide Web)	
WYSIWYG (What You See Is What You Get)	
XML (eXtensible Markup Language)	
XPS (XML Paper Specification)	
XSLT (eXtensible Stylesheet Language Transformation)	
ZB (Zetta Byte)	

Reference

[1] 孙洁,李一,湛邵斌.实用 IT 英语[M].2 版.北京:人民邮电出版社,2014.

[2] 朱龙,刘长君.计算机专业英语[M].2 版.北京:人民邮电出版社,2018.8.

[3] 刘兆毓,郑家农.计算机英语实用教程[M].5 版.北京:清华大学出版社,2019.

[4] 吴云丽,于萍萍.计算机专业英语[M].北京:北京邮电大学出版社,2020.

[5] 张强华,司爱侠.人工智能专业英语[M].北京:人民邮电出版社,2021.

[6] 司爱侠,张强华.计算机英语[M].4 版.北京:人民邮电出版社,2022.

[7] Timothy J. O'Leary, Linda I. O'Leary, Daniel A. O'Leary.计算机科学引论[M].北京:机械工业出版社,2018.

[8] Email Basics [EB/OL]. [2022-01-10]. https://edu. gcfglobal. org/en/email101/introduction-to-email/1/.

[9] Basic Computer Skills:how-to-set-up-a-wifi network[EB/OL].[2022-01-15].https://edu.gcfglobal.org/en/basic-computer-skills/how-to-set-up-a-wifi-network/1/.

[10] Structured data & unstructured data[EB/OL].[2022-01-15]. https://www.talend.com/resources/structured-vs-unstructured-data/.